STRESS AND EPIGENETICS IN SUICIDE

STRESS AND EPIGENETICS IN SUICIDE

VSEVOLOD ROZANOV

Odessa, Ukraine

ACADEMIC PRESS

An imprint of Elsevier
elsevier.com

Academic Press is an imprint of Elsevier
125 London Wall, London EC2Y 5AS, United Kingdom
525 B Street, Suite 1800, San Diego, CA 92101-4495, United States
50 Hampshire Street, 5th Floor, Cambridge, MA 02139, United States
The Boulevard, Langford Lane, Kidlington, Oxford OX5 1GB, United Kingdom

Notices
Knowledge and best practice in this field are constantly changing. As new research and experience broaden our understanding, changes in research methods, professional practices, or medical treatment may become necessary.

Practitioners and researchers must always rely on their own experience and knowledge in evaluating and using any information, methods, compounds, or experiments described herein. In using such information or methods they should be mindful of their own safety and the safety of others, including parties for whom they have a professional responsibility.

To the fullest extent of the law, neither the Publisher nor the authors, contributors, or editors, assume any liability for any injury and/or damage to persons or property as a matter of products liability, negligence or otherwise, or from any use or operation of any methods, products, instructions, or ideas contained in the material herein.

Library of Congress Cataloging-in-Publication Data
A catalog record for this book is available from the Library of Congress

British Library Cataloguing-in-Publication Data
A catalogue record for this book is available from the British Library

ISBN: 978-0-12-805199-3

For information on all Academic Press publications
visit our website at https://www.elsevier.com/books-and-journals

  Working together
to grow libraries in
developing countries

www.elsevier.com • www.bookaid.org

Publisher: Mara Conner
Acquisition Editor: April Farr
Editorial Project Manager: Timothy Bennett
Production Project Manager: Sue Jakeman
Designer: Miles Hitchen

Typeset by Thomson Digital

Contents

4. Biological Embedding—How Early Life Stress Shapes Behaviors Later in Life and How Vulnerability is Built

5. Interactions and Integrations—Biobehavioral Model of Suicide Based on Genetics, Epigenetics, and Behavioral Adjustment

6. Ideas for Prevention

Introduction

Recent observations show that despite many positive changes in the last century (i.e., better nutrition, medical and social aid, education, technological progress, etc.) there is a paradoxical increase of mental health problems in the younger generation, including impaired problem solving, risky behaviors, addictions, and suicide. The last is the subject of deep concern of the society, especially parents, school staff, army, which is meeting more difficulties with newcomers' screening, and universities. The processes that make younger generation so vulnerable have been developing for a comparatively short (even in historical terms) period of time-starting from 1980s of the last century and until today. Among the set of factors that underlie it may be biological, psychological, social, and existential, involving meanings and purposes that youngsters see in their lives. There are nevertheless many pieces of evidence that the leading cause of accumulation of adolescents' problems is the psychosocial stress, a complex feeling, which is based on modernization, inequalities, and instability and is multiplied by information technologies. All phenomena that are discussed are complex and multifaceted, linked together, and interdependent. There is a great need for integration of knowledge in this field, which may serve the goal of better understanding adolescents' suicides and development of more integrated and effective preventive strategies.

One of the most promising fields of knowledge that potentially may serve the goals of integration of heterogeneous factors of suicide today is behavioral epigenetics, especially in combination with the neurobiology of stress response and neuroplasticity. Epigenetics is understood as changes in genes activity due to chemical modifications of chromatin (or some other mechanisms, for instance, involving noncoding RNAs) without touching the sequence of DNA itself. Animal studies and observations in humans testify that epigenetic transformations are driven by different environmental stimuli that are crucial for adaptation and survival in the changing environment. For human beings, social environment is the most important in this sense. Exact mechanisms are not studied yet; nevertheless, there is a consensus that epigenetic marks are triggered by internal regulators and biologically active molecules, including those involved in stress response. The last is extremely important. This puts together stressful exposures of critical life periods of the child and behaviors of young parents, dysfunctional parenting strategies, and their transgenerational progression. In this book, several possible models of suicidal behavior, which are based on the central role of epigenetic modifications and which incorporate early and later life stress, psychological factors like perceived stress and coping abilities, social aspects (psychosocial stress), and existential and cognitive factors will be presented and discussed.

The book is not a systematic review or analysis of the state of the art in the field, but more an attempt to integrate knowledge of the interdisciplinary nature. It raises more questions that give answers, and is aimed to attract the attention of researchers and to raise a wider discussion in this field. A better understanding of interactions of epigenetics and social, psychological, and existential factors united by stress mechanisms and their input in suicide propensity may promote new approaches to suicide prevention. Understanding of the role of stress as of internal psychic process and links of these experiences to values, meanings, and purposes in life may be helpful in individual and group measures to prevent future suicides in young people.

In accordance with this general understanding, the book consists of six chapters. First, we are presenting results of the objective analysis of the situation with suicides among young people on the global level. Further discussion is concentrated on modern psychosocial stress and its biological mechanisms, interrelations between social factors and biological mechanisms, understanding how typical stress response can be triggered by purely psychological states that are closely related to social factors. The further discussion concentrates on the modern understanding of epigenetic mechanisms and their role in programming behaviors and traits important for adaptation. Main attention is paid to the so-called biological embedding-mechanism of incubation of vulnerability that may become obvious later in life and manifests itself in different diseases, disorders, and maladaptive behaviors. In subsequent chapters, this modeling is applied directly to suicidal behavior, providing a scheme of possible interactions between biological, psychological, social, and existential factors during development of the personality and considering intergenerational transmission and bidirectional loops between epigenome, personality, perceptions, and social environment. The model finally becomes a source of outlining main directions of preventive strategies aimed to break off a vicious circle.

Recent Tendencies in Suicide and Mental Health Among Younger Generations and Current Explanations

Suicide is a serious problem of public health in the modern world. It becomes an even more serious problem with global information pressure and universal availability of information today, including descriptions of suicide cases by the mass media. Of course, suicide mortality is much lower than mortality from most widespread noncommunicable diseases like myocardial infarction or cerebral stroke, and this is often used as justification for lower interest and fewer efforts regarding prevention. But the psychological and moral impact of suicide is very high and cannot be measured in purely economic or organizational terms. It is especially true for youth suicides which constitute a growing problem in many countries on all continents. Of course, if look at the death rates in demographic groups, suicide rate in the older age (>75) population is the highest, while in people under 24 it is the lowest. On the other hand, for the last 60 years there is a distinct global shift—among those who die of suicide the portion of young people under 45 exceeded the portion of people over 45 years old (Bertolote & Fleischmann, 2009). Moreover, suicides in the youngest, for instance aged 15–24 years, become more and more prevalent. Official statistics of WHO says that in the developed world in 2010 suicide was one of the leading causes of death (and in some countries–the first) for people aged 15–49 years (Institute for Health Metrics and Evaluation, 2010). Thus, we are facing a grave paradox—better living conditions, better and more balanced nutrition, modern medical aid with all possible progress in psychiatry and psychology, better education, technological progress, and free access to information, as well as rich personal freedoms and higher value of human life in the modern society—that does not prevent young people

Stress and Epigenetics in Suicide. http://dx.doi.org/10.1016/B978-0-12-805199-3.00001-4

from killing themselves. On the contrary, it seems that all these progressive developments have sometimes negative consequences for some part of new generations of young people, while vulnerability of the youngest, aged 13–15 years, seems to be the highest. Of course, it raises a lot of questions and concerns in the public. Those who have devoted themselves to suicide research and prevention are often approached by mass media and representatives of public with the most frequent question—why? What is the reason? And of course, who is to blame?

Adolescents and young adults' suicides have been in the focus of many studies, which are systematically overviewed (Pfeffer, 2000; de Wilde, 2000; Apter, Bursztein, Bertolote, Fleischman, & Wasserman, 2009a; Rozanov, 2014b; Kolves & DeLeo, 2014, 2015; Apter & Gvion, 2016). Age and sex differences, cultural peculiarities, life stress (negative life events), mental illness, and drug abuse, as well as mass media exposure and contagion, are most widely discussed. Among social and psychological factors, role of family situation, school problems, risky behaviors, bullying and peer victimization, nonsuicidal self-harm, and some psychopathologies and traits (anxiety, depression, sensation and novelty seeking, low self-esteem, and thwarted identity) are mentioned. Much attention is paid to neurodevelopmental mechanisms and such predisposing factors as the differential maturation of emotional and cognitive structures of the brain, which may result in risky behaviors, mood swings, impaired decision making, problem-solving, and cognitive rigidity (Pfeffer, 2000; de Wilde, 2000; Apter, Krispin, & Bursztein, 2009b; Rozanov, 2014b; Kolves & DeLeo, 2014, 2015; Apter & Gvion, 2016). All these factors will be discussed throughout the book, although here we aim to concentrate on temporal changes of suicide rates in adolescents as the sign of magnitude and evolution of the problem.

Apparently, adolescent and young people suicides are rather rare with regard to absolute figures and even rates, as calculated to 100,000 of corresponding age (de Wilde, 2000). On the contrary, suicide of a young person produces the most oppressive effect on close surrounding and on the society in general so far as it is more often reported by mass media and is accompanied by greater social resonance. Suicide of the young is the most well-noticed and tragic event easily provoking in the society a feeling of grave tendencies that contradict all our expectations and perceptions of the youth. Possibly, this is the reason why such influential organization as WHO in 2001, that is, 15 years ago, have published in its Bulletin a short paper under provocative title, *Choosing to Die—A Growing Epidemic Among the Young* (Brown, 2001). After ascertaining disturbing facts about the rise of suicides in the younger generation, the author cites the opinions of the leading western suicidologists (José Bertolote, Diego De Leo, Danuta Wasserman, and Morton Silverman) concerning the reasons and explanations for this phenomenon. Proposed reasons are as follows: the fast changes that affect different spheres of life in modern societies, existential problems of the young, erosions of traditions, problems in relationship of generations, increased consumption of alcohol and drugs, and growing prevalence of mental disorders among young people. All this makes the younger generation more vulnerable according to the cited article. This article (and the title itself) is very typical for the general representation of youth suicide in printed and electronic sources. Although WHO specialists in the well-known suicide prevention recommendations for mass media warn against the use of such wording as "suicide epidemic," emotions seem to prevail and existing journalists stamps are too strong. That's why it is really important to look at the problem from the point of view of facts and evidence-based publications. On the contrary, one must be aware that scientific publications are influenced by contextual factors too and authors' perceptions and personal feelings toward the problem may have some effect.

YOUTH SUICIDES—IS IT REALLY GROWING WORLDWIDE?

The situation with youth suicides needs detailed analysis because in different countries one can come across both alarmist articles and counter-messages that are trying to sooth emotional evaluations. Is the situation really so grave? What are the temporal trends of youth suicides? Did they really start to grow quite recently, or did this tendency start 50–60 years ago and have already a long history? What are the tendencies worldwide? Does it happen in Europe, or in Americas, what are the tendencies in the post-Soviet world, in the growing China and India, and in other parts of the world? Is it an isolated trend or is it linked to changing pattern of mental health of youngsters? And the most important—what explanations are suggested today by professional and expert communities and the wide public, except already mentioned?

We will try to elucidate these issues and to find answers to above-listed questions in this chapter before we start to discuss the role of epigenetics and stress in suicide. We deliberately focus on suicides in the youngest so far as new generations seem to be more vulnerable, which may have an explanation within the concept of epigenetics and stress and which will be addressed later.

In this chapter, we are focused on completed suicides, not suicide attempts or suicidal ideation. A subject of the analysis in this regard is the mortality statistics over time. It simplifies a task because it "cuts of" a set of other suicide manifestations (attempt, thought, etc.) which can be revealed only during focused study, and which can produce very ambiguous results due to complexities of inquiries, ways of asking questions, level of emotional contact between the investigator and type of surveyed contingent (de Wilde, 2000). A fatal case is an indisputable event, and comparison of national mortality statistics is supposed to give a more accurate result, more or less free of subjective opinions. However, there is still a problem of registration of causes of death. The existing taboos concerning suicides: religious and family traditions have a great influence on official statistics too. It is especially characteristic of the so-called "developing" countries (as a rule, it is the countries with traditional, differing from globalized), structure of society, more traditional values, and so on. Thus, it is necessary to be cautious with some data too. Nevertheless, "big figures" are perceived with more confidence, especially when these figures are monitored over time during historically distinct periods, or are covering vast territories and hundred million populations.

It is also methodologically important to differentiate accurately age groups if we are pursuing the goal to build an objective picture of this problem. It is known that within a short period of maturation the suicide prevalence among young people changes sharply. Indexes of suicides in the age group of 5–14 years are really very low (approximately 0.5–2.5 per 100,000 depending on the country or the region); however, among adolescents and young adults (14–25 years) these are already 10 times higher (5.0–28.0 per 100,000) (Apter et al., 2009b). Cases of death, for instance from hanging among children under 10–11 years, are considered accidents though there are always doubts remaining. Thus, a very short time, a year or two, may be crucial, while statistics usually covers bigger age intervals. Most of the authors who are involved in pediatric suicidology agree that children's suicides are generally underestimated and statistical information on them is always below real values (Apter et al., 2009b; Rutz & Wasserman, 2004; Wasserman, Cheng, & Jiang, 2005). All this gives an opinion about context against which this problem is discussed here.

To obtain objective picture free from emotional evaluations, it is advisable to look at the problem from the geographical perspective, and starting with generalizing publications, which are trying to build a worldwide perspective. In the last decades, suicides among children and teenagers in a global context were analyzed in details in the article of Pelkonen and Marttunen (2003) and in the special review prepared for the authoritative collective monograph which appeared in Oxford University Press (Apter et al., 2009a). Both reviews contain all official statistics of suicides, available by the time of publication, among adolescents and young people, unfortunately within not coinciding age ranges, but in comparative aspect, over the countries and continents. All studies are based on official WHO statistics which is lagging behind for 2–3 years from the real situation.

The main conclusion which can be made of the extensive data set is the following. Indexes of suicides among teenagers correlate with indexes in the general population due to which ranking of various countries by suicide rate in young people remains almost invariable. So, in full accordance with the well-known facts, first place on teenage suicides in 1997–98 belongs to Russia, Kazakhstan, Estonia and Lithuania, and also Finland, Slovenia, Belarus, and Ukraine. On the other pole, there are countries with the lowest levels, such as Armenia, Azerbaijan, Greece, Macedonia, and Portugal. At the same time, among the countries with the highest level, there is New Zealand (Pelkonen & Marttunen, 2003). In another review reflecting situation for the beginning of the year 2000 and covering a wider set of countries, first-ranking positions are also occupied by Lithuania, Russia, Finland, and New Zealand, while on another pole there are Portugal, Greece, and Egypt. At the same time, on the very first place with an incredibly high level of youth suicides there is Sri Lanka (Apter et al., 2009a). Both publications are representing a comparative (between countries and continents) situation for the last available year, thus they have not much value from the point of view of the analysis of changing trends. On the contrary, in both publications there are indications of growing of suicides among children and adolescents in specific countries and regions, including Eastern Europe, South America, South Africa, and Asia. Safer picture is observed in Western Europe and North America though some local peaks are still evident (Apter et al., 2009a).

One of the attempts to describe adolescents' suicides global time trends, not only to reflect a static picture, was done by Wasserman et al. (2005). Suicide rates among adolescents in the age group of 15–19 years, from 90 countries out of 130 WHO member states, were analyzed. In addition, data from 26 countries for the period 1965–99 were examined on the basis of WHO Mortality Database. This set of countries included 18 states of Western and Central Europe and another 8 countries from outside Europe (Mauritius, Canada, USA, Hong Kong, Singapore, Japan, Australia, and New Zealand). The trend of suicide rates was estimated on the basis of descriptive analysis. Authors have concluded that there is a steady rise of suicides in young males for the whole observed period in the majority of countries. It was especially clear in the states outside Europe where it was particularly marked before 1980. This publication and some special notices from reviews of Pelkonen and Marttunen (2003) and Apter et al. (2009a) give an impression that in many places in the world there was a clear rising trend in suicides from the period 1960s to the end of 1990s of the last century. On the contrary, a report on aggregated rates of suicides in young people aged 15–19 years (without differentiation between sexes) of OECD countries (which includes not only almost the same set of economically developed countries but also Israel, Turkey, Poland, and some others) testifies of lowering trend of youth suicides for the period of 1990–2009 (McLoughlin, Gould, & Malone, 2015). This gives an impression how different may be general conclusions in cases of different sets of parameters, information sources, and historical periods covered. The longer is the period of observation, the better so far as most recent trends may depict only part of the tendency. It must also be taken into consideration that OECD report includes most economically developed countries where suicide prevention measures are more often supported on the government level and where the general level of public safety and availability of medical aid is the highest in the world.

In a quite recent publication of Kolves and DeLeo (2015), trends for of suicides in young people for 2 decades covering the end of 20th century and beginning of the 21st century are presented. Authors have analyzed suicide rates in adolescents aged 15–19 years between 1990 and 2009 worldwide on the basis of suicide mortality data obtained from WHO sources. In total, 81 countries or territories, having data at least for 5 years in 1990–99 and in 2000–09, were included in the analysis. Additional analysis of regional trends in 57 countries was performed. The authors have been focused only on statistically verified results. Their analysis revealed a very subtle but distinct declining trend in the global suicide rate in young people. More significant decline (approximately 17%) was found for males and less pronounced decline (14%) for females in Europe. On the contrary, there was registered a significant increase in South American countries for males (55%), and a little less pronounced rise for females (42%). Although other world regions did not show significant changes, there were several significant peaks in different countries. The authors explain the decrease in Western countries by overall improvements in global health, while increases in several South American countries are thought to be related to the economic recession. In another paper of the same authors, global situation in the younger age group (10–14 years) is described. It was found that global rate in boys in 81 countries has shown a minor decrease, while the rate in girls, in contrast, has shown a minor increase. At the regional level, rise in younger boys in the post-Soviet space is mentioned (Kolves & DeLeo, 2014). Thus, when looking at global data, fast and steady growth from 1960 to 1990 seems to have slowed and even reversed for males, while there is still growth in the youngest females. The positive trend in Western Europe, USA, and Canada is counterbalanced by negative changes in Latin America and post-Soviet space.

The approach based on national statistics is not free from shortcomings. In many countries, ethnic and cultural heterogeneity is the reason of differing regional levels of suicides. In can be found in many country reports, and as we will be able to see further, very often higher suicide rates belong to ethnics and groups which are living according to old traditions and are confronted by modernization and stress associated with it. We will briefly look at studies which are giving evidence of time change of adolescents' suicides for the last decades, covering as a long period of observation as possible, on the more local level, within the individual states, and, where possible, on the regional level. Such approach is especially relevant for countries built on the federative basis, like Russian Federation.

SUICIDE AMONG YOUNG PEOPLE IN THE EUROPEAN CONTINENT

In 2004, analysis of mortality of teenagers (15–19 years) from suicides in the countries of the European region according to the WHO classification (the countries of the European continent including all countries of the former USSR, Turkey, and Israel) from 1979 to 1996 was published (Rutz & Wasserman, 2004). In 21 of 30 countries, which were in the list, for the specified period the levels of suicides among youngest men have increased, while among men above 20 years and in older groups this tendency was much less expressed and even decrease of levels was observed. The most significant growth was observed in Belarus, Kazakhstan, and Ireland. Among young women growth of suicides was observed only in 18 countries, and it was not so expressed, except for essential rise in Ireland and Norway (Rutz & Wasserman, 2004). These data convincingly testify of a tendency of growth of adolescents' suicides in the European region in the last decades of the 20th century.

Studies in separate European countries add substantially to these data and confirm the above-mentioned conclusions. In Finland, data on suicide rates among young people up to 18 years of age for the period from 1969 to 2008 are published. Levels of suicides in males grew from 1969 to 1989, while among females during this period no essential changes were observed. Since the beginning of the 1990s, levels among young men started to decrease, while the growth of indexes among girls became even more obvious. All violent suicides decreased for males and increased for females for the whole period of study, the fact that violent suicides in females are growing is pointed by authors as an alarming and nontypical tendency (Lahti, Räsänen, Riala, Keränen, & Hakko, 2011).

The growth of children and adolescent suicides is observed for a number of years in Sweden where suicide prevention measures (including at the state level) are already implemented for decades. Teenagers of northern regions of Sweden, where mortality from the external reasons in general is higher, appeared to be the most vulnerable part of the population (Johansson, Stenlund, Lindqvist, & Eriksson, 2005). The highest rates are found among native peoples of the North—reindeer-herding Laplanders or Saami people, representatives of the Finno-Ugric group (Ahlm, Hassler, Sjölander, & Eriksson, 2010).

In the Great Britain, levels of suicides among young people recently are higher than in other age groups—suicides are the leading cause of death of people aged 15–19 years. On the contrary, during the period from 1997 to 2010 teenage suicides are decreasing. As in many other countries of Europe, in Great Britain young people aged 15–19 years demonstrate much higher suicide mortality than adolescents aged 10–14 years—among young males suicide mortality is 3–4 times higher than among females (Windfuhr et al., 2013). Recent statistical reports evidence on the growth of suicides among adolescents aged 10–14 years and young people aged 15–19 years in the United Kingdom—this is viewed as a tendency which should be confirmed (Stallard, 2016).

In Austria, mortality from suicides of adolescents under 14 for the period from 1970 to 2001 was analyzed. For the 30 years, the level of suicides decreased among boys, while among girls it remained almost invariable. Authors emphasize that decrease of indexes of suicides among male teenagers coincides with the general tendency among the Austrian population (Dervic et al., 2006). The resistance of suicide rates in young females deserves special attention from this point of view, as it is confirmed in many other studies that come from economically developed and stable countries.

Italy is the country with a low level of suicides (especially in the southern regions); nevertheless, suicides of teenagers are attracting a lot of attention. In one of the publications, authors analyzed mortality of teenagers aged 10–17 years for the period from 1971 to 2003. Remarkably, mortality of teenagers from all external causes decreased, except for suicides, which remained almost constants. Authors consider that suicide belongs to the causes of death that are resistant to wide prevention measures among teenagers that are applied recently (Pompili, Vich, De Leo, Pfeffer, & Girardi, 2012). Another work covering almost the same period of time, but using another source of information, claims insignificant rise of suicide indexes—the greatest rise was registered in young males aged 15–19 years (Campi, Barbato, D'Avanzo, Guaiana, & Bonati, 2009). Thus, in Italy, on the background of very low levels of suicides in the general population, the same dynamics of adolescents' suicides can be seen—an increase or stable level, while there is a decrease of this indicator in the older age groups.

The situation in Eastern and Southern Europe has similar tendencies. In Poland, significant increase of suicides in 1999–2007 was registered among the youngest girls aged 10–14 years, while in boys aged 10–14 and 15–19 years and in girls of the senior group there was no distinct change. On the contrary, negative correlation between suicides and group poisonings and undetermined deaths in older groups found in the study may indicate hidden suicides (Kułaga, Napieralska, Gurzkowska, & Grajda, 2010). Thus, there is a distinct growth of suicides among the youngest females in Poland, which has been registered also in some other European countries.

Confirming evidence is presented in the study from Bosnia and Herzegovina. Authors analyzed suicides among young people during 1986–90 and 2002–06, before and after the so-called Bosnian war of 1992–95. Research showed that levels of suicides (especially with the use of firearms) elevated in the postwar period, mainly among children aged 14 years and younger, while among girls and 15–19 years old boys these have decreased (Fajkic et al., 2010). Article from Croatia also reports a rise of suicides among teenagers and specifically mention the significant growth of cases with the use of firearms (Dodig-Curkovic, Curkovic, Radic, Degmecic, & Filekovic, 2010). These publications clearly indicate the role of war crisis as a stressor in the society and higher availability of firearms as means of suicide associated with it, especially in relation to teenagers.

In Southern Europe and Asia Minor levels of suicides are traditionally low. It is the case for teenagers and young people in Epirus—the northwest agricultural region of Greece, known for its most ancient history—where no growth of suicidality among the youth is registered. The reason of this positive tendency Greek authors see first of all in the traditional way of life in this patriarchal region, where strong family traditions remain a natural protective factor for the youth (Vougiouklakis, Tsiligianni, & Boumba, 2009). Report from geographically close Turkey is not so optimistic—a rise of teenage suicides is mentioned (Arslan, Akcan, Hilal, Batuk, & Cekin, 2007; Uzun et al., 2009).

Presented data give an impression that on the European continent, which is one of the most economically developed parts of the world, suicides in adolescents and young adults are characterized by certain problems, but in general the situation is not critical and remains under control. For the most economically developed (OECD) countries from 1990 to 2009, there was a slow lowering of suicide rates in teenagers. On the contrary, this positive trend only recently has replaced rising trend that lasted from 1970s to 1990s. Existing sex and ethnic variations in suicidality are influenced by cultural, historical, psychological, relational, and socioeconomic factors (McLoughlin et al., 2015). One of the trends noticed by many authors is growing of suicides in the younger females. We are mostly interested in socioeconomic factors so far as they greatly contribute to the existing level of psychosocial stress in the given society. In view of this, it is, therefore, appropriate to consider the situation in the countries of the former USSR which have undergone painful transition period associated with economic and social problems.

CHILDREN AND ADOLESCENT SUICIDES IN THE COUNTRIES OF THE FORMER USSR

The situation in the countries of the former USSR deserves separate consideration. Initial decrease and the further sharp rise of suicides in these countries in the period from 1985 to 2000 were repeatedly discussed from different points of view, but mainly with regard to such factor as alcohol consumption (Värnik, Wasserman, Dankowicz, & Eklund, 1998; Mäkinen, 2000; Nemtsov, 2003). This is an important factor, but of course not a decisive one, especially if all society is taken into consideration.

We have presented alternative view (supported by quasiexperimental data) that one of the leading reasons for substantial lowering of suicides during "perestroika" (reconstruction, a process of democratization started by Gorbachev in 1986) was social optimism and positive expectations, while further dramatic rise in suicide mortality in 1992–2002 was the result of severe psychosocial stress caused by disintegration of the USSR and following socioeconomic problems. Alcohol factor explains only part of suicides mostly related to mental health problems, while existential factors, associated with socioeconomic transition and disintegration of the general social and cultural space, may have a wider impact, involving bigger contingents with and without definite mental health problems.

The fall of the USSR was associated with a complex of severe and long-lasting negative social, economic, and political processes that have hit huge groups of people and touched different ages, from youngsters, who were still dependent on the economic state of their caregivers, to retired citizens, mostly dependent on social welfare (Rozanov, 2014a). This period was also characterized by introduction of new values, harsh confrontation with realities of the emerging capitalism, destruction of the existing social stereotypes and habitual ways of life, huge increase of a social inequality and rude competitiveness, loss of social cohesion, disintegration of the existing system of medical and social aid, and many other negative consequences. All these factors may have contributed to dramatic suicides rise in the countries of former USSR, while the rise of alcohol consumption probably reflected a traditional way of adaptation to stress (Rozanov, 2014b). On the other hand, disparities between former Soviet Republics are characteristic.

Shortly after the fall of the USSR in the period from 1995 to 2002, there was a considerable rise in suicides in all industrialized post-Soviet countries with initially high suicide level (Russia, Latvia, Lithuania, Estonia, Belarus, Ukraine, and Kazakhstan), after which a gradual decrease started. Recently, the majority of the countries, which appeared on the world map after the collapse of the USSR, are on the lowering trend regarding suicide indexes; nevertheless, they keep their "places" in the world rank of suicidality, which depicts "cultural resistance" of this indicator. In the countries of the Caucasus, Transcaucasia, and Central Asia, focused on agriculture, with the traditional structure of a family and society and with initially low suicide rates, fluctuations of the levels of suicides for the specified period were much less marked and suicide indexes remained almost steadily low (Rozanov, 2014a). This is an important background which helps to understand adolescents' suicides in the countries of the former USSR.

The situation with teenagers' suicides in the Russian Federation was thoroughly analyzed in the detailed report, initiated by UNICEF

(Ivanova et al., 2011). According to the data collected in this source, the suicide rate among adolescents aged 15–19 years in the Soviet Union started to grow from 1965 to the beginning of the 1970s, reaching a peak of the entire Soviet period of history. Then there was a gradual decline, which became especially clear during the democratization period (1986–89), but soon in the early 1990s it was replaced by a sharp rise, which was observed until 2002, after which again started a steady decline, which in recent years is not so convincing (Ivanova et al., 2011). This evolution, in general, coincides with the fluctuations of the rate of suicide in the general population (Rozanov, 2014a), suggesting that youth suicides largely depend on the same global factors as adult suicides but with some quantitative differences.

In the last decade, suicides among adolescents are the highest in Kazakhstan, Russia, Lithuania, and Estonia, despite the fact that the last two countries belong to the European Union where ideology and practice of suicide prevention have a long history. In all these countries, adolescent suicides in the last 10 years are declining, though remain the highest in the WHO European Region. This decline does not fully coincide with the general population trends. In Russia, for example, between 1999 and 2009, mortality from suicides among adults on the background of economic development has decreased by approximately 30%, while among teenagers it has only decreased by 10% (Rozanov, 2014b). Analysis of the situation on the local level (a million city of Odessa) in the period from 2002 to 2010 has revealed a growing, not lowering, trend of suicides in the young people under 18 (Rozanov, Valiev, Zakharov, Zhuzhulenko, & Kryvda, 2012).

It is possible to hypothesize that differences in adolescents suicides in new independent states have roots in the cultural traditions (stabilizing factor) and in modernization that is spreading on the post-Soviet space (destabilizing factor). The highest rate of teen suicide is reported recently in Kazakhstan, the country where traditions, patriarchal relations, and high birth rate are neighboring with huge economic development, westernized modernization, and wide representation of information technologies (UNICEF, 2012). In Kazakhstan, a lot of suicides occur not only in the northern industrial regions, where environmental pollution and proximity to the Semipalatinsk nuclear testing site may have an influence but also in the southern regions, with the dominating Muslim population and very traditional values.

Another confirmation of the role of misbalance between traditions and modernity comes from the Russian Federation where national districts and regions with "unprecedented high levels of suicide among adolescents" were identified. This, in particular, may be seen in the Republic of Tyva, as well as the republics of Buryatia and Altai (Polozhy, Kuular, & Dukten-ool, 2014). According to Russian authors, the highest teenage suicide indexes are found among above-mentioned ethnics and northern indigenous peoples, in particular, the Yakuts, who belong to Turkic ethnic and language group, and Evenks, also known as Tungus. The reasons from one side are seen among such factors as high alcohol consumption, inadequate social infrastructure, and poor access to mental health resources (Polozhy et al., 2014). There is also another point of view based on the idea that the natural way of life and traditions of these peoples are coming into conflict with the rapidly changing technological and social environment, which causes constant psychosocial stress. This idea is expressed in the study of Tsyrempilov, in which the author, involving the concept of ethnogenesis developed by Lev Gumilyov, suggests low "passionarity" (spiritual way characterized by low competitiveness, more reserved, a restricted level of claims, mild contemplation instead of strive for achievement, etc.) of circumpolar nations. Such mentality is seen as one of the main reasons, among others, why these nations and especially young people are facing difficulties in adaptation to the rapid changes in the modern environment (Tsyrempilov, 2012).

Another reason may be associated with more measurable factors. Studies on the peculiarities of the cognitive processes of the Yakuts and Evenks provide evidence that living for centuries in severe conditions of the North with sensory deprivation has caused in these ethnics specific features of information processing and other cognitive peculiarities. A modern social environment with fast changes, urbanization, and information overload comes into conflict with these peculiarities which may be the reason for the higher level of psychosocial stress. It may not be perceived as stress on the personality level, but may undermine psychological stability and mental health, especially of youngsters (Semenova, 2013). Other authors also consider the conflict between traditional and modern cultures and constant necessity to adapt to new conditions as main reasons of high suicide rate in the youth of northern ethnics like Nenets, Itelmen, Koryak, Chukchi, Yukaghir, and so on in Russian Federation national autonomies (Lyubov, Sumarokov, & Konoplenko, 2015).

Thus, over the past 30 years, the countries of the former Soviet Union provide an example of sharp fluctuations of young people suicide rates, which largely coincide with changes in suicides in the general population but which also have unique trends in several regions. It may be seen in Kazakhstan and in northern and Siberian national districts of the Russian Federation, where suicide rates among adolescents are very high and where national traditions, ways of life, and rather archaic values preserved for centuries are confronted by liberal economic development, westernized traditions, individualism, and competitiveness associated with frustrations and stress. It presumably hits the most vulnerable part of the young who try to adapt to a new reality with information technologies and high pace of living.

YOUTH SUICIDES IN ASIA AND THE FAR EAST

Moving further to the East, the situation in China, India, and East Asia should be considered. These regions in the postwar years have faced rapid economic and social changes (South Korea, Hong Kong, Malaysia, the Philippines, and Japan). In recent decades, due to the transition to market relations, there was a rapid economic and technological development in such countries as China and Vietnam. Thus, modernization has affected this region at full scale, and many countries with very specific cultural traditions have apprehended western values and became impregnated with them, youth being on the frontline of this conflict of traditions and cultures. Based on available data, the level of teenage suicides in China is substantially lower than in Europe (Apter et al., 2009a). On the contrary, the youth mortality rate due to external causes is increasing in recent decades, and these changes are more pronounced in rural areas (Jiang et al., 2011; Liu et al., 2012). It is noteworthy that in Hong Kong, which has a longer history of economic development and the western way of life, in the age group of 15–24 years suicide mortality rates is much higher than in mainland China (Apter et al., 2009a). Moreover, for the period from 1980 to 2003 in Hong Kong there has been a steady increase in suicides among young people aged 10–24 years (Shek, Lee, & Chow, 2005). A study from South Korea also informs about sharp rise of suicides among young people aged 15–24 years in the period from 2001 to 2011—from 9.0 to 16.1 per 100,000 in males and from 6.2 to 15.0 in females (Park, Im, & Ratcliff, 2014). Authors of these studies are suggesting a well-developed theory about the reason for this alarming tendency—a conflict between traditional Confucian cultural values and more liberal sociocultural norms and meanings exported from western society and apprehended mostly by youth. This societal psychosocial stress is accompanied by increasing media coverage of suicides, which is considered another most important factor of youth suicide growth. Park et al. (2014) consider that media suicide exposure may cause changes in attitudes to suicidal acts, which may lead to more accepting thinking about suicide among the young. This conceptual understanding also explains higher suicide rates in rural areas and among women in China, in both cases, the conflict and "cultural ambiguity" (pressure of parents who are bearing traditions and influence of the modern culture which suggests much more liberal attitudes) reaches the maximum (Park et al., 2014). Here it is appropriate to mention that in China completed suicides in rural areas and among young females are very common, suicides in females outscoring suicides in males, which contradict European and American tendencies and western culture in general (McLoughlin et al., 2015).

Rather common picture is seen in India where the study in Vellore (South East Region) revealed that suicide rates among adolescents aged

10–19 years reached an incredibly high level—152 per 100,000 among women and 69 per 100,000 among young men, females obviously out-reaching males (Aaron et al., 2004). As was already mentioned, very high suicide rates in young people are registered recently in Sri Lanka, in particular, in the age group of 15–19 years among men these reach 59 per 100,000, while among women these reach 42 per 100,000 (Apter et al., 2009a). In the Philippines, the analysis of the dynamics of suicides from 1974 to 2005 has been published. Indexes of suicides, previously very low (approximately 0.2 per 100,000 in men and 0.1 per 100,000 in women), rose more than 10 times between 1984 and 2005. When considering different age groups, the most intensive growth was among women aged 15–24 years (Redaniel, Lebanan-Dalida, & Gunnell, 2011).

Summing up we can conclude that growing suicides in the younger generation are noted also in Southeast Asia and in India. Studies in these regions are not so numerous and convincing and time trends are often not very clear if compared with the Western Europe or Northern America, where suicides in young people are scrutinized for decades. However, existing studies testify about negative tendencies, and general opinions and emotional tone of publications reveal concerns of authors regarding this negative development. Among factors contributing to the high suicide rate in this very economically active region of the world, there has been the role conflict between collectivism and individualism, the rigidity of hierarchical social structure in the face of happening changes, which makes stress even more acute. With regard to youth suicides, such factors are also mentioned as the strengthening of the repressive element in the education system, and most importantly a conflict between the local and foreign culture, promoted by the media (Kim & Singh, 2004).

SITUATION IN NORTH AMERICA, AUSTRALIA, AND NEW ZEALAND

The United States and Canada are the most economically developed countries, which in the cultural context are very close to the European (Euro-Atlantic) civilization. It is, therefore, interesting to compare youth suicide trends in this world region with Western Europe. Suicides among young people aged 15–19 years in the United States began to grow rapidly since the beginning of the 1960s of the last century with an increase by 250% by the end of the 1980s, followed by a further slight decline in 1990s. As a result, over 40 years to the beginning of the 20th century, suicide rates in this age group have doubled (Beck-Little, 2011). If we look at the youngest age groups in the United States (15–19 and 15–24 years old) in comparison with older groups over the last 40–50 years, the picture is very confirmatory—an increase of suicides among the youngest against the decline in the grown-ups groups (McKeown, Cuffe, & Schulz, 2006). Since the 1990s, however, teenage suicide is mainly declining, although from time to time short peaks are registered. The American Society is ethnically very heterogeneous; in this regard, most of the studies evaluate the differences between different racial groups. Both in adults and adolescents, suicide rates are highest among whites, markedly lower among Hispanics and even lower among African Americans. However, the highest levels in adolescents are registered among the indigenous population of the American Indians, suicide rates among them is 3.5 times higher than the national average (Greudanus, 2007).

According to a study from Canada, from 1980 to 2008 among young people aged 10–14 and 15–19 years there is a gradual decrease in the level of suicide by approximately 1% a year. However, more precise analysis by sex and age revealed an important feature—this decrease was observed only among boys, while in the girls there was an increase of 50% over the mentioned period (Kirmayer, 2012; Skinner & McFaull, 2012). Although suicides with the use of firearms decreased, the number of suicides by asphyxia has increased sharply. The authors consider that measures aimed to reduce the availability of weapons (firearms in particular) were effective, but widening of use of asphyxia to achieve euphoria has led to a new wave of deaths (Kirmayer, 2012; Skinner & McFaull, 2012). These observations draw our attention to the importance of the availability of means of suicide, as well as the "fashion trends" among teenagers as factors that influence both suicide rates and age and sex differences. These fashion trends show that teenagers are very dependent on their reference groups and are often unable to resist the pressure that comes from their closest environment. In Canada, very high suicide rates are registered among indigenous Inuit people, especially in the young Inuit (Webster, 2016).

Teenage suicides in Australia and New Zealand—the two island states, formed mostly by colonists from the United Kingdom and Ireland—also provide important information. One can say that these countries currently have a pretty high level of economic development and enjoy social stability, good level of social welfare, and environmental well-being. Not surprisingly, the rates of suicide in Australia in the general population are low and comparatively stable. However, a completely different picture is observed among youth. During the 40-year period from the mid-1960s of last century, suicide rate among young people aged 15–24 years has increased almost 3 times (Cantor, Neulinger, & De Leo, 1999). This increase is very similar to the growth in the United States, which allows assuming that certain similarities exist between the psychosocial processes in these countries, which are so geographically distant but culturally are rather close. There is also a distinct difference between Australian indigenous people and white population—suicide rate in aborigines is 2–3 times higher than among the descendants of white colonists, and this is especially obvious for youth suicides (Cantor et al., 1999). Quite similar processes are observed in New Zealand. The level of suicides in men aged 15–24 years started to grow in the 1970s of the last century and has increased significantly in the mid-1990s and only then began to decline. It is significant that among the native population (Maori) suicide rates are higher than among white Australians (Beatrais, 2003; Beautrais and Fergusson, 2006). In conclusion, adolescents' suicides in the "new world" largely are showing the same evolution—steady rise from 1980s to the end of the century with subsequent slow decrease and periodical peaks. Quite expectedly indigenous youth are at higher risk. If we think in terms of adaptation to the changing social environment, it may be said that the first wave of problems that have emerged due to complexities of adaptation in these countries have already passed. It possibly came earlier than in the countries that have emerged after the fall of the socialist system, including China and Vietnam.

SUICIDE AMONG YOUNG PEOPLE IN LATIN AMERICA AND AFRICA

In this region of the world, levels of suicides are generally low. However, the available sources specify that the tendency to a growth of suicidality among youth touched this region too. For instance, in Argentina growth of suicides among young men of 20–24 age groups is registered quite recently—in 2005–07 (Bella et al., 2013). In Chile, indexes of suicides among young men and women almost doubled from 1995 to 2003 (Apter et al., 2009a). In Brazil, the level of suicides in the general population is low, but on this background, there is significant growth in suicides among youth. In particular, from 1980 to 2000 suicide level increased in age group of 15–24 years almost 20 (!) times. These data are confirmed in several types of research; growth among women was lower but still definite—4 times (Apter et al., 2009a). In Mexico, national level is also rather low (5 per 100,000); however, in young people since the beginning of the 1990s up to the beginning of the 2000s there was almost triple growth of suicides among women and double growth among men aged 11–19 years (Apter et al., 2009a). It is remarkable that the highest level of suicides in Mexico is noted in regions of traditional residence of the most ancient Maya people—states Yucatan and Campeche (Baquedano, 2009). The main reason for this phenomenon is seen in the stressful conflict of the traditional religions which are still determining cultural life of the population (though they are coexisting with Catholicism) and modern domination of consumerism, alcohol and drugs abuse, social and economic deprivation, as well as low educational level of the teenagers (Baquedano, 2009).

Due to various reasons, objective data on suicides at the national level in African region are fragmentary. We can be guided by recent reviews of the studies on the African continent, which state that suicide is a growing problem of public health on the continent, including youngsters (Schlebousch, Burrows, & Vawda, 2009; Mars, Burrows, Hjelmeland, & Gunnell, 2014). Main reasons of growing suicides are seen among such factors as social disintegration and uncertainty regarding personal future, including increasing competitiveness and stress (Schlebousch et al., 2009). In another study, Flischer, Liang, Laubscher, and Lombard (2004) reported a significant increase in suicides among youth in Africa between 1968 and 1990, especially marked in males. Among risk factors, substance abuse and family dissolution are mentioned (Flischer et al., 2004).

YOUTH IS UNDER THREAT—PRELIMINARY CONCLUSIONS

Results of our analysis of the situation with youth suicides worldwide are leading to rather definite conclusions. For the last several decades, suicides among teenagers and youth aged 11–24 years are growing, and it is definitely a global tendency. It is confirmed by a large number of studies and publications which are based on national and local statistics and are covering big time intervals. Growing suicides in young people are inherent to both economically well-off countries of Europe and North America, which had a long-lasting period of stable development, and for the countries of the former Soviet Union, which endured numerous socioeconomic difficulties for the last 3 decades, and also for Asia and the Pacific and Latin America. In some countries, this rise started in the 1960s of the last century and was persisting until the end of 1990s, while today these tendencies are not so noticeable. This can be seen in Western European countries—Great Britain, Austria, Finland, Switzerland, and Italy. It seems that the peak of teenage suicides has also passed in the USA, Canada, Australia, and New Zealand. In other countries like South Korea, Taiwan, and Sri Lanka, countries of the Latin America, and in the post-Soviet space, rising of youth suicides is noticeable in the very last decades. Due to existing time-shifts between periods of rising and lowering of suicides in different parts of the world, global evaluations are not very conclusive. However, many publications support our analysis (Wasserman et al., 2005; Apter et al., 2009b; Kolves & DeLeo, 2014, 2015). It is necessary to stress that in many publications it is mentioned that in traditional high-risk groups (middle and older age males) suicide rates are not growing or even descending, while younger generations are showing growing rates. In general, the world rising trend of suicide has slowed down (Bertolote & De Leo, 2012), but this is not true for the young people.

One of the most important conclusions is that despite the fact that in many European countries and USA, Australia, and New Zealand new prevention programs with special emphasis on youth were introduced and implemented, it was not possible to lower youth suicides to the level that existed 50–60 years ago. Moreover, in many countries rise of violent suicides in very young females is reported, while male suicides remain stable or even decrease, quite possibly due to an introduction of preventive strategies.

The complex and mixed picture of growing and lowering suicide trends in adolescents may be related to many psychosocial factors, including alcohol and drug use fluctuations, bullying and peer victimization, prevalence of mental health problems like depression or anxiety, and many others (Orbach & Iohan-Barak, 2009; Wasserman et al., 2012; McLoughlin et al., 2015). Suicide is often a consequence of mental disorder or mental health problems, thus the association with these tendencies is quite understandable. On the contrary, mental health problems and suicides evolution are derivatives from more global stress factors, which widely depict changing trends of the most general social environment. Social environments bring a variety of influences and acting factors, which can lead to both positive and negative psychosocial consequences, depending on the context and the group of the population that may benefit or suffer from these influences. Among them are changing socioeconomic situation, periods of development and recession from one side, and perceived or real social inequalities from another side. As it was extensively discussed by Wilkinson and Pickett (2009), different psychosocial problems in the society, including mental health, criminality, and conflicts, are related to existing inequalities, while the general economic wealth of the society has much less impact. These authors are proposing status syndrome as a main provocative factor that translates inequalities into social tension and promotes conflicts, frustration, and dissatisfaction with life in the whole society, from rich to poor (Wilkinson & Pickett, 2009). Very close ideas are discussed by Curtis, Curtis, and Fleet (2013) from New Zealand. These authors studied suicide rates in different age groups and have revealed consequent rising trends of suicides in each next 10-year cohort. They consider it a consequence of economic downturn in New Zealand and the reaction of each cohort when these individuals reach adulthood and feel the impact of poverty and social comparisons that inevitably arise (Curtis et al., 2013). We may add to this that this factor may have even greater impact in societies that have just joined the global context, like post-Soviet countries, where there was a dramatic rise in inequalities, and this ugly picture became very evident to youth through information pressure and quick shift of status syndrome from "parent level" to "teenagers level." The last was often due to a quick erosion of values of collectivism that were promoted for many years in these countries and which suddenly lost their relevance and attractiveness, giving the way to individualism, competitiveness, and status evaluation.

If among "newcomers" quick and painful transition to global world produced very strong but comparatively short-time rise in suicides, after which some adaptation started, small indigenous ethnics seem to be in the situation of constant pressure for the whole period of industrialization. In a large part of publications, authors specify that young representatives of the autochthonic ethnics appear to be the most vulnerable part of the population with regard to suicidality. This can be seen among Laplanders in Sweden, Tyvinians, Yakuts, Evenks, Nenets, Buryats, and Altay peoples in Russia, Maori in New Zealand, North American Indians, Inuit people in Canada, and the Australian aborigines. In many papers, a reasonable point of view is expressed that threat for the cultural identity of indigenous peoples in the modern and globalized world may be the main reason of their elevated suicidality. In the last decades, these peoples need to adapt even more because of growing pressure of lifestyles dictated by the western culture, which promotes such values as consumerism, competitiveness, individualism, and materialism. It comes into conflict with traditional values and religions of these ethnics while young generation appears in the situation "between two fires"—traditions, supported by the families and lures, promoted by the external world. Often their economic situation is lower than that in their country as a whole and it may make cultural conflict even more acute. Elevated rates of suicidality in indigenous people could be related to loss of family links and community support, socioeconomic difficulties, marginalization, racism, loss of religious affiliation, inequality in education, cultural clashes with parents, thwarted hopes and lack of belonging, historically conditioned loss of native lands, dismantling of cultural processes and structures (McLoughlin et al., 2015). While discussing Inuit youth M. Kral (2013) points also to such factors as modern gender ideologies and new cultural models of love and sexuality, which add to the conflict between traditional relations and novelty. One of the probable reasons, why this conflict becomes so obvious quite recently, is the influence of the modern information technologies which have enhanced their pressure and multiplied information flows falling upon the youth, making risks more evident. Considering usual psychological difficulties which are experienced by the teenager or the young individual in the course of the personality growth, socialization, solving existential questions, and search of his/her own place in the world, this external pressure may become a serious threat to identity and integrity, especially for autochthonous minorities youth.

So far as many authors who study this subject point on cultural shifts and social or economic processes as main factors of youth suicides, it is important to stress that many other risks, such as mental disorders associated with suicide, like depression, anxiety, and addictions are also to a great extent provoked by the same wide factors. Various mental health problems known as most important predictors of suicidal behavior are of complex bio-psychosocial origin and are also largely dependent on cultural and social context. From this point of view, it is necessary to point that mental health problems are also growing among the young people worldwide and that this may add to growing suicide rates. Moreover, some psychological traits that are associated with suicidal behavior, for instance neuroticism, though largely dependent of genetic predispositions, are the subject of slow evolution in the modern world. This will be discussed further.

YOUTH MENTAL HEALTH PROBLEMS—ARE THEY GROWING TOO?

Problems of mental health in children and adolescents are diverse and not at all rare. There is a great body of evidence that though some suicides may occur on the background of the absence of any disorder, in a great portion of cases suicide is a consequence or complication of a disturbed mental state. Affective disorders, anxiety, substance abuse, eating disorders, and personality disorders, as well as more serious mental illnesses, are known risk factors of suicide (Wasserman et al., 2012). Many objective studies directly prove that high suicide rates in certain populations are coinciding with the high prevalence of mental disorders. For instance, in Tyva, where suicides in adolescents are considered the highest in the world, the prevalence of psychopathologies among boys is also substantially higher than in general population (Slobodskaya & Semenova, 2015). The same can be said about indigenous population of Canada, USA, Australia, and New Zealand (Clifford, Doran, & Tsey, 2013). Such examples are numerous: suicides and mental health problems go together, mental health problems and more serious disorders like clinically diagnosed depression or bipolar disorder are most important predictors of suicide, while completed suicides rate and suicide attempts rate are often seen as indicators of the mental health of the population (Wasserman, 2016). Socioeconomic problems and conflict of cultures, seemingly so subtle but actually so destructive regarding their consequences for social cohesion, may be the reason for both.

It is important to mention that disturbed mental conditions, which are most prevalent in the adolescent populations (depressive symptoms, anxiety, substance abuse, adjustment disorders), are not referred as severe; on the contrary, they are comparatively mild. On the contrary, their interpersonal and social impact may be high. It is, therefore, important to differentiate between different types of disturbances and problems of mental health in youngsters. Mental health means much more than just absence of mental illness or disorder; nevertheless, it is often evaluated on the population level as the prevalence of diagnosable mental illnesses and disorders. From this point of view, positive and negative mental health are often understood as two sides of the same coin (Keyes, 2005). Such approach is especially relevant when speaking about the wide range of disturbances which remain under threshold and do not meet all diagnostic criteria. They are often referred as mental health problems (BMA Board of Science, 2006). These dynamic states are the main reason for everyday problems of the youth; they often lead to difficulties in interpersonal relationships, psychological development, the capacity for learning, and in ability to build resilience toward distress. During adolescence, all these kinds of problems may interact with stress and complex feelings that are experienced by the young personality while he/she forms the perception and understanding of the world, struggles for the place among the peers, and starts to build relations with an opposite sex. In the case of negative development and conflicts accumulation, in the case of loss of contacts with the older generation, which are very probable in the situation of cultural (traditions vs. modernity) conflict, these problems may be exacerbated. When these problems become persistent and severe and start to affect functioning on a day-to-day basis, and if strong sides of the personality (referred as positive mental health) cannot counterbalance it, they may be diagnosed as mental health disorders (BMA Board of Science, 2006). Of course, serious mental illnesses like schizophrenia may also emerge in adolescence, but they remain rather rare and their prevalence is more or less stable.

From our perspective it is important to mention that most widely spread problems of mental health in adolescents and young people, such as depression, anxiety, substance abuse, conduct, adjustment, personality, and eating disorders, are often associated with suicidal behavior, while depression, anxiety, and substance abuse are known as most serious risk factors for completed suicide (Orbach & Iohan-Barak, 2009; Apter et al., 2009b). It is, therefore, important to see if the prevalence of mental health problems in young people shows the same pattern for the last decades as suicides.

Epidemiological studies and clinical observations evaluate that globally approximately 20% of adolescents have some mental health problems of different severity. Special studies conducted in Europe give estimates within 14–23% depending on age group and sex (Belfer, 2008; Ravens-Sieberer, Erhart, Gosch, & Wille, 2008). Another study provides data that for the last year a quarter of teenagers may have mental health problems and a one-third in a lifetime. Anxiety disorders are the most frequent condition in children, followed by behavior disorders, mood disorders, and substance abuse (Merikangas, Nakamura, & Kessler, 2009). This data give the impression of the magnitude of the problem of children mental health disturbances in a modern world.

The question we are mostly interested in is whether we observe parallel processes—suicides and mental health problems prevalence growth among teenagers in a historical perspective. Suicides are better registered, while the prevalence of mental health problems and disorders can be evaluated only within focused epidemiological studies and using special criteria. There is always a concern that development of child psychiatry, an introduction of more structured diagnostic criteria and higher awareness in the medical community (both among psychiatrists and GPs or pediatrics, who are recently involved in children mental health evaluation), may lead to an increase of registered diagnoses. There is also a lot of concern about the pressure of pharmaceutical companies lobbying medication for depression and anxiety, which in turn promotes more enthusiastic diagnostic approaches in the medical community. Thus, one of the serious concerns is that changing prevalence of mental health problems may be the consequence of hyperdiagnostics.

We would like to stress that many professional publications contradict this point of view: rise of mental health problems is registered globally and in the young generations in particular (Mitikhina, Mitikhin, Iastrebov, & Limankin, 2011; Steel et al., 2014; Rozanov, 2015). One of the recent reviews dedicated to this problem draws the following conclusion: in the Western world since the beginning of the 20th century to the end of the 1990s sharp rise of various manifestations of anxiety and depression among children and teenagers is observed (Twenge, 2011). The author emphasizes that in the American society at the beginning of the 20th century depression was diagnosed in 1–2% of the population, while since the end of the 1960s its frequency grew 10 times and by the end of the 1990s some studies claimed that almost half of population was captured by depression. Almost all available evidences suggest a sharp rise in anxiety, depression, and other mental health problems among western youth between the early 20th century and the 1990s. Since the end of the 1990s, possibly due to the introduction of new antidepressants, a number of suicides stabilized or even lowered, while psychosomatic complaints and feeling of being overwhelmed continued to increase (Twenge, 2011).

These observations are supported by the metaanalysis of the psychological profiles of college students obtained by a widely used and standardized personality test MMPI in a wide historical range—between 1938 and 2007. This analysis has revealed an increase in several clinical scales, suggesting a certain evolution of personality traits and accumulation of psychopathologies (many of which may be related or could be considered as indicators of Eysenck's dimension of neuroticism) over time (Twenge et al., 2010). The last conclusion may have serious consequences with regard to suicidality, so far it is well known that high neuroticism is one of the most often found psychological traits in suicide attempters (Williams & Pollock, 2000; Rozanov & Mid'ko, 2011). Neuroticism in adolescents as a personality trait is related to objectively measured stress reactivity (Evans et al., 2016). Evidence of accumulation of psychopathologies and evolution of personality traits, especially neuroticism, in a historically short period of time is of great interest. Neuroticism has been viewed as a stable, genetically based trait, so changes that occur throughout the lifespan and between age cohorts may mean more complicated and dynamic control of this trait (Barlow, Ellard, Sauer-Zavala, Bullis, & Carl, 2014).

Similar results are presented in the publication of S. Callishaw and co-authors from Great Britain. The study assessed the evolution of the conduct, hyperactive, and emotional problems over a 25-year period in three general population samples of UK adolescents. Moreover, it was supported by comparable questionnaires completed by parents of 15–16 year olds at each time point (1974, 1986, and 1999). Results of the study showed that for the mentioned period significant growth of conduct problems among teenagers was observed (approximately by 1.5 times). It has affected both males and females, all social classes and all family types. There was also evidence for a rise in emotional problems and hyperactive behavior. What is really important is that the authors have provided evidence that observed trends were unaffected by possible changes in reporting mode (Collishaw, Maughan, Goodman, & Pickles, 2004).

Concerns regarding youth mental health and suicides have led to several international research projects aimed to evaluate the extent of the problem and to suggest solutions for prevention. One of the most recent studies in 11 European countries has revealed high level of risk behaviors among adolescents aged 14–16 years, approximately one-third of pupils experienced subthreshold depression, 8% of the sample was categorized as seriously depressed, 23.3% of pupils experienced subthreshold anxiety, and 4.7% of pupils reported severe-to-extreme anxiety, while conduct problems occurred in 10.3% of the sample. One of the most important results from this study is the identification of several subgroups inside this big and international sample. Due to special statistical analysis, three groups of adolescents were identified: a low-risk group (57.8%) including pupils with low or very low frequency of risk behaviors; a high-risk group (13.2%) including pupils who scored high on all risk behaviors, and a third group ("invisible" risk, 29%) including pupils who were positive for high use of Internet/TV/video games, sedentary behavior, and reduced sleep. Pupils in the "invisible" risk group had an almost similar prevalence of suicidal thoughts, anxiety, and depression as children from the high-risk group. The prevalence of suicide attempts was 10.1% in the high-risk group, 5.9% in the "invisible" group, and 1.7% in the low-risk group (Carli et al., 2014). Thus, almost half of more than 12,000 of students had some problems of mental health. There is no comparison over time in this study, but the magnitude of the problems of adolescents seems to accumulate, especially with emerging Internet addiction.

In overall, there is a strong evidence, not only qualitative estimations, that mental health problems are growing in the young people for the last decades, and this coincides to a certain extent with the growth of suicides among representatives of this age group. We have recently reviewed tendencies in mental health in the general population and in adolescents in different countries and came to the conclusion that accumulation of mental health problems and growing prevalence of disorders are reported in many countries and continents, so it is also a global trend (Rozanov, 2015). This coincides with other authors' analysis (Collins et al., 2011).

SUBJECTIVE COMPLAINTS IN ADOLESCENTS AND VALUES SHIFT AS SIGNS OF STRESS

When we imagine adolescents we usually perceive them as healthy, joyful, full of life and energy, and in the modern world —more technologically competent than older generation, especially regarding Internet technologies. That is true, but only to a certain extent and in relation only to the portion of them. All above-presented data suggest that on the subjective level modern youngsters (and especially the most vulnerable ones) suffer psychologically much more than their counterparts 50–60 years ago. When subjective complaints of children and teenagers are evaluated, a picture of diverse negative feelings appears—feeling of being overwhelmed, experiencing stress, anxiety, having symptoms of depression, sleeping problems and disturbed eating behavior, symptoms of Internet or social networks addiction, and abuse of psychoactive agents, including cigarettes and alcohol. Richard Eckersley in his remarkable paper provides results of a survey of Australian students, which revealed that great proportion of young people aged 17–18 years felt lonely (18%), felt hopeless and depressed (20%), were stressed (31%), were irritated and worried to much (35 and 42%), and had difficulties calming down when upset (48%). In America, similar study revealed that for the last 2 weeks 54% of students had felt overwhelmed by all they had to do, 50% had felt exhausted, 19% had felt severe anxiety, 17% had felt that things were hopeless, 12% had felt overwhelming anger, and 10% were so depressed that it was difficult to function (Eckersley, 2011).

Survey of the senior teenagers (16–18 years, total number 1027) in Sweden showed that the fatigue and grief were felt by approximately 40% of girls and 20% of boys, while the feeling of constant tension due to high requirements at school was reported by 64% of girls and 39% of young men (Wiklund, Malmgren-Olsson, Ohman, Bergström, & Fjellman-Wiklund, 2012). Mental health, stress, and subjective complaints in students are the growing problems not only for schools but also for colleges and universities. The intensity of educational process, information overloads, and an unhealthy lifestyle are mentioned as possible reasons (Storrie, Ahern, & Tuckett, 2010). In our survey of adolescents belonging to different types of schools, from elite gymnasiums to normal district and boarding schools, we revealed that problems of mental health (signs of depression, anxiety, perceived stress, and hopelessness) were strongly associated with loss of meaning in life, a complex feeling which in turn is linked to more severe suicidal thoughts and attempts, suggesting that "existential vacuum" is inherent in teenagers and plays an important role of a risk factor (Rozanov, Rakhimkulova, & Ukhanova, 2014). It was also noticed that the higher is the status of the school, the more problems are experienced by the students and the more subjective complaints are expressed.

Twenge (2011) when trying to outline primary reasons for generational differences in mental health pays much attention also to such factors as values and meanings. She is speaking about "cultural shifts" that became evident in the western society in the second half of the last century. She is arguing that rise of psychopathologies and mental health problems in youth is linked with transition from "intrinsic" values and goals (aimed to meet such human needs as competence, affiliation, and autonomy) to "extrinsic" ones, which are associated with external rewards like status, money, and recognition (Twenge, 2011). Eckersley points that such shifts are translated on the personal level into the overall pressure of the false values of success instead of personal development and emotional maturation, in the necessity to meet high and often unrealistic expectations with subsequent feeling of being trapped and a higher risk of failure and goal conflict (Eckersley, 2011). He also points to results of surveys of life goals among American college students which showed that the biggest shift since 1970s and 1980s was between "developing a meaningful philosophy of life" and "being very well of financially" (Eckersley, 2011).

Subjective feelings and complaints together with values shifts give an impression of the level of internal psychological (psycho-emotional, mental) stress experienced by young people in the modern society. The origin of this stress is associated with the lifestyle in modern urbanized and globalized world, based on values of success, individualism, materialism, and competitiveness. It may be said that this, in turn, is based on the existing neoliberal economic model, which ensures constant pressure to consume more and introduce sophisticated methods of making people want more than they need. There is no doubt that modern information technologies and ubiquity of Internet content make this pressure even more profound. It may be said that the modern world is turning into a "global village," where primitive needs are easily satisfied, while critical thinking is often not supported. Infantile ideas are very easily distributed by the Internet and social networks and frustration cover constantly increasing contingents, especially when real life comes into conflict with nonrealistic expectations. Moreover, social networks are dividing people into small groups, where erosion of "big meanings" becomes even more probable. All this adds to general context and "cuts the grass from under feet" of the younger generation which appears to be more vulnerable than adults.

The picture drawn gives an impression how sex and ethnic variations in suicidality are embedded within cultural, historical, psychological, and socioeconomic domains. We can fully agree with McLoughlin et al. (2015) that it is vital to adopt a holistic approach that incorporates an awareness of all factors that underlie suicidal behavior. It is our deep conviction that stress is one of the central factors deeply interconnected with all the above-mentioned domains. The complex picture of existential, psychological, and emotional problems experienced by most vulnerable adolescents (it is important to remember that only part, though growing, of youngster actually suffer from these problems) of course means a high level of stress experienced by the younger generation. We are going to discuss the essence, biopsychological mechanisms, peculiarities, and consequences of this stress in the next chapter. It is also very important to understand better the complex bio-psycho-social nature of the vulnerability that makes a portion of the young population more susceptible to stress and more prone to self-harm and, ultimately, to suicide. Growing suicides in the younger people, growing prevalence of mental health problems, and more serious clinically evident disorders, as well as an evolution of some personality traits, subjective complaints, and values shifts, implicate complex interplay between stress, vulnerabilities, personality, social, cognitive, and existential factors. Recent development in the neurobiology of stress and understanding genetic and epigenetic basis of stress vulnerability and resilience give a chance to discuss these interplay on a more substantial and objective level.

References

Aaron, R., Joseph, A., Abraham, S., Muliyil, J., George, K., Prasad, J., Minz, S., Abraham, V. J., & Bose, A. (2004). Suicides in young people in rural southern India. *Lancet, 363*, 1117–1118.

Ahlm, K., Hassler, S., Sjölander, P., & Eriksson, A. (2010). Unnatural deaths in reindeer-herding Sami families in Sweden, 1961–2001. *International Journal of Circumpolar Health, 69*(2), 129–137.

Apter, A., & Gvion, Y. (2016). Adolescent suicide and attempted suicide. In D. Wasserman (Ed.), *Suicide. An unnecessary death* (2nd ed., pp. 197–213). New York: Oxford University Press.

Apter, A., Bursztein, C., Bertolote, J., Fleischman, A., & Wasserman, D. (2009a). Suicide in all continents in the young. In D. Wasserman, & C. Wasserman (Eds.), *Oxford textbook on suicidology and suicide prevention. A global perspective* (pp. 621–628). New York: Oxford University Press.

Apter, A., Krispin, O., & Bursztein, C. (2009b). Psychiatric disorders in suicide and suicide attempters. In D. Wasserman, & C. Wasserman (Eds.), *Oxford textbook on suicidology and suicide prevention. A global perspective* (pp. 653–660). New York: Oxford University Press.

Arslan, M., Akcan, R., Hilal, A., Batuk, H., & Cekin, N. (2007). Suicide among children and adolescents: data from Cukurova, Turkey. *Child Psychiatry and Human Development, 38*, 271–277.

Baquedano, G. (2009). Maya religion and traditions influencing suicide prevention in contemporary Mexico. In D. Wasserman, & C. Wasserman (Eds.), *Oxford textbook on suicidology and suicide prevention. A global perspective* (pp. 77–84). New York: Oxford University Press.

Barlow, D. H., Ellard, K. K., Sauer-Zavala, S., Bullis, J. R., & Carl, J. R. (2014). The origins of neuroticism. *Perspectives on Psychological Sciences, 9*, 481–496.

Beatrais, A. L. (2003). Suicide in New Zealand: time trends and epidemiology. *Journal of the New Zealand Medical Association, 116*, 1175.

Beautrais, A. L., & Fergusson, D. M. (2006). Indigenous suicide in New Zealand. *Archives of Suicide Research, 10*, 159–168.

Beck-Little, R. (2011). Child and adolescent suicide in the United States: a population at risk. *Journal of Emergency Nursing, 37*, 587–589.

Belfer, M. L. (2008). Child and adolescent mental disorders: the magnitude of the problem across the globe. *Journal of Child Psychology and Psychiatry, 49*, 226–236.

Bella, M. E., Acosta, L., Villace, B., Lopez de Neira, M., Enders, J., & Fernandez, R. (2013). Analysis of mortality from suicide in children, adolescents and youth. Argentina, 2005–2007. *Archivos Argentinos de Pediatria, 111*, 16–21.

Bertolote, J. M., & De Leo, D. (2012). Global suicide mortality rates—a light at the end of the tunnel? *Crisis, 33*, 249–253.

Bertolote, J. M., & Fleischmann, A. (2009). A global perspective oon the magnitude of suicide mortality. In D. Wasserman, & C. Wasserman (Eds.), *Oxford textbook of suicidology and suicide prevention: a global perspective* (pp. 91–98). New York: Oxford University Press.

BMA Board of Science. (2006). *Child and adolescent mental health. A guide for healthcare professionals*. London: British Medical Association.

Brown, P. (2001). Choosing to die—a growing epidemic among the young. *Bulletin of the World Health Organization, 79*, 1175–1177.

Campi, R., Barbato, A., D'Avanzo, B., Guaiana, G., & Bonati, M. (2009). Suicide in Italian children and adolescents. *Journal of Affective Disorders, 113*, 291–295.

Cantor, C., Neulinger, K., & De Leo, D. (1999). Australian suicide trends 1964–1997: youth and beyond? *Medical Journal of Australia, 171*, 137–141.

Carli, V., Hoven, C. V., Wasserman, C., Chiesa, F., Guffanti, G., Sarchiapone, M., Apter, A., Balazs, A., Brunner, R., Corcoran, P., Cosman, D., Haring, C., Iosue, M., Kaess, M., Kahn, J. -P., Keeley, H., Postuvan, V., Saiz, P., Varnik, A., & Wasserman, D. (2014). A newly identified group of adolescents at "invisible" risk for psychopathology and suicidal behavior: findings from the SEYLE study. *World Psychiatry, 13*, 78–86.

Clifford, A. C., Doran, C. M., & Tsey, K. (2013). A systematic review of suicide prevention interventions targeting indigenous peoples in Australia, United States, Canada and New Zealand. *BMC Public Health, 13*, 463.

Collins, P. Y., Patel, V., Joestl, S. S., March, D., Insel, T. R., Daar, A. S., Scientific Advisory Board, the Executive Committee of the Grand Challenges on Global Mental Health, Anderson, W., Dhansay, M. A., Phillips, A., Shurin, S., Walport, M., Ewart, W., Savill, S. J., Bordin, I. A., Costello, E. J., Durkin, M., Fairburn, C., Glass, R. I., Hall, W., Huang, Y., Hyman, S. E., Jamison, K., Kaaya, S., Kapur, S., Kleinman, A., Ogunniyi, A., Otero-Ojeda, A., Poo, M. M., Ravindranath, V., Sahakian, B. J., Saxena, S., Singer, P. A., & Stein, D. J. (2011). Grand challenges in global mental health. *Nature, 475*, 27–30.

Collishaw, S., Maughan, B., Goodman, R., & Pickles, A. (2004). Time trends in adolescent mental health. *Journal of Child Psychology and Psychiatry, 45*, 1350–1362.

Curtis, B., Curtis, C., & Fleet, R. W. (2013). Socio-economic factors of suicide—the importance of inequality. *New Zealand Sociology, 28*, 77–92.

Dervic, K., Friedrich, E., Oquendo, M. A., Voracek, M., Friedrich, M. H., & Sonneck, G. (2006). Suicide in Austrian children and young adolescents aged 14 and younger. *European Child and Adolescent Psychiatry, 15*, 427–434.

Dodig-Curkovic, K., Curkovic, M., Radic, J., Degmecic, D., & Filekovic, P. (2010). Suicidal behavior and suicide among children and adolescents—risk factors and epidemiological characteristics. *Collegium Antropologicum, 34*, 771–777.

Eckersley, R. (2011). Troubled youth: an island of misery in an ocean of happiness, or the tip of an iceberg of suffering? *Early Intervention in Psychiatry, 5*, 6–11.

Evans, B. E., Stam, J., Huizink, A. C., Willemen, A. M., Westenberg, P. M., Branje, S., Meeus, W., Koot, H. M., & van Lier, P. A. (2016). Neuroticism and extraversion in relation to physiological stress reactivity during adolescence. *Biological Psychology, 117*, 67–79.

Fajkic, A., Lepara, O., Voracek, M., Kapusta, N. D., Niederkrotenthaler, T., Amiri, L., Sonneck, G., & Dervic, K. (2010). Child and adolescents suicides in Bosnia and Herzegovina before and after the war (1992–1995). *Crisis, 31*, 160–164.

Flischer, A. J., Liang, H., Laubscher, L., & Lombard, C. (2004). Suicide trends in South Africa, 1968–1990. *Scandinavian Journal of Public Health, 32*, 411–418.

Greudanus, D. E. (2007). Suicide in children and adolescents. *Primary Care, 34*, 259–273.

Institute for Health Metrics and Evaluation. (2010). *Global burden of disease: Generating evidence, guiding policy*. Seattle, WA: Institute for Health Metrics and Evaluation.

Ivanova, A. E., Sabgayda, T. P., Semyonova, V. G., Antonova, O. I., Nikitina, O. Yu., Evdokushina, G. N., & Chernobavskiy, M. V. (2011). *Smertnost rossiyskih podrostkov ot samoubiystv (Suicide mortality among Russian adolescents)*. Moscow: UNICEF.

Jiang, G., Choi, B. C., Wang, D., Zhang, H., Zheng, W., Wu, T., & Chang, G. (2011). Leading causes of death from injury and poisoning by age, sex and urban/rural areas in Tianjin, China 1999–2006. *Injury, 42*, 501–506.

Johansson, L., Stenlund, H., Lindqvist, P., & Eriksson, A. (2005). A survey of teenager unnatural deaths in northern Sweden 1981–2000. *Accident Analysis and Prevention, 37*, 253–258.

Kral, M. J. (2013). "The weight on our shoulders is too much, and we are falling": suicide among Inuit male youth in Nunavut, Canada. *Medical Anthropology Quarterly, 27*, 63–83.

Keyes, C. L. M. (2005). Mental illness and/or mental health? Investigating axioms of the complete state model of health. *Journal of Consulting and Clinical Psychology, 73*, 539–548.

Kim, W. J., & Singh, T. (2004). Trends and dynamics of youth suicides in the developing countries. *Lancet, 363*, 1090–1091.

Kirmayer, L. J. (2012). Changing pattern in suicide among young people. *Canadian Medical Association Journal, 184*, 1015–1016.

Kolves, K., & DeLeo, D. (2014). Suicide rates in children aged 10–14 years worldwide. *British Journal of Psychiatry, 205*, 283–285.

Kolves, K., & DeLeo, D. (2016). Adolescent suicide rates between 1990 and 2009: analysis of age group 15–19 years worldwide. *Journal of Adolescents Health, 58*(1), 69–77.

Kułaga, Z., Napieralska, E., Gurzkowska, B., & Grajda, A. (2010). Trends in children and adolescents deaths due to suicide, event of undetermined intent and poisoning in Poland in the years 1999–2007. *Przeglad Eepidemiologiczny, 64*, 551–556.

Lahti, A., Räsänen, P., Riala, K., Keränen, S., & Hakko, H. (2011). Youth suicide trends in Finland, 1969–2008. *Journal of Child Psychology and Psychiatry, 52*, 984–991.

Liu, Q., Zhang, L., Li, J., Zuo, D., Kong, D., Shen, X., Guo, Y., & Zhang, Q. (2012). The gap in injury mortality rates between urban and rural residents of Hubei Province, China. *BMC Public Health, 12*, 180.

Lyubov, E. B., Sumarokov, Y. A., & Konoplenko, E. R. (2015). Zhiznestoykost' i factory riska suitsidalnogo povedeniya korennyh malochislennyh narodov severa Rossii (Resilience and suicide behavior risk factors in indigenous peoples of the Russian North). *Suitsidologia (Suicidology), 6*, 23–30.

Mäkinen, I. H. (2000). Eastern European transition and suicide mortality. *Social Science and Medicine, 51*, 1405–1420.

Mars, B., Burrows, S., Hjelmeland, H., & Gunnell, D. (2014). Suicidal behaviour across the African continent: a review of the literature. *BMC Public Health, 14*, 606.

McKeown, R. E., Cuffe, S. P., & Schulz, R. M. (2006). US suicide rates by age groups, 1970–2002: an examination of recent trends. *American Journal of Public Health, 96*, 1744–1751.

McLoughlin, A. B., Gould, M. S., & Malone, K. M. (2015). Global trends in teenage suicide: 2003–2014. *QJM: An International Journal of Medicine, 108*, 765–780.

Merikangas, K. R., Nakamura, E. F., & Kessler, R. C. (2009). Epidemiology of mental disorders in children and adolescents. *Dialogues in Clinical Neuroscience, 11*, 7–20.

Mitikhina, I. A., Mitikhin, V. G., Iastrebov, V. S., & Limankin, O. V. (2011). Psychicheskoe zdorovye naseleniya mira: epidemiologicheskiy aspect (zarubeshniye issledovaniya 2000-2010 gg) (Mental health of the world population: epidemiological aspects (the analysis of foreign research results for 2000–2010). *Zhurnal Nevrologii I Psychiatrii imeni S.S. Korsakova (Korsakov Journal of Neurology and Psychiatry), 111*, 4–14.

Nemtsov, A. V. (2003). Suicide and alcohol consumption in Russia, 1965–1999. *Drug and Alcohol Dependence, 71,* 161–168.

Orbach, I., & Iohan-Barak, M. (2009). Psychopathology and risk factors for suicide in the young. In D. Wasserman, & C. Wasserman (Eds.), *Oxford textbook on suicidology and suicide prevention. A global perspective* (pp. 634–641). New York: Oxford University Press.

Park, B. C., Im, J. S., & Ratcliff, K. S. (2014). Rising youth suicide and changing cultural context in South Korea. *Crisis, 35,* 102–109.

Pelkonen, M., & Marttunen, M. (2003). Child and adolescent suicide. Epidemiology, risk factors, and approaches to prevention. *Pediatric Drugs, 5,* 243–265.

Pfeffer, C. R. (2000). Suicidal behavior in children: an emphasis on developmental influences. In K. Hawton, & K. van Heeringen (Eds.), *The international handbook of suicide and attempted suicide* (pp. 237–248). Chichester: John Wiley & Sons.

Polozhy, B. S., Kuular, L. Y., & Dukten-ool, S. M. (2014). Osobennosti suitsidal'noy situatsii v regionah s sverhvysokoy chastotoy samoubiystv (na primere respubliki Tyva) (Peculiarities of suicidal situation in the regions with ultrahigh suicide rate (on an example of the Republic of Tyva). *Suitsidologia (Suicidology), 1,* 11–18.

Pompili, M., Vich, M., De Leo, D., Pfeffer, C., & Girardi, P. (2012). A longitudinal epidemiological comparison of suicide and other causes of death in Italian children and adolescents. *European Child and Adolescent Psychiatry, 21,* 111–121.

Ravens-Sieberer, U., Erhart, M., Gosch, A., & Wille, N. European Kidscreen Group. (2008). Mental health of children and adolescents in 12 European countries – results from the European KIDSCREEN study. *Clinical Psychology and Psychotherapy, 15,* 154–163.

Redaniel, M. T., Lebanan-Dalida, M. A., & Gunnell, D. (2011). Suicide in the Philippines: time trend analysis (1974–2005) and literature review. *BMC Public Health, 11,* 536.

Rozanov, V. A. (2014a). Psychosocial stress and suicidal behavior. In U. Kumar (Ed.), *Suicidal behavior. Underlying dynamics* (pp. 80–97). New York, London: Routledge.

Rozanov, V. A. (2014b). Suitsidy sredy podrostkov—chto proishodit i v chem prichina? (Suicides in children and adolescents—what is happening and what may be the reason?). *Suitsidologia (Suicidology), 5,* 16–31.

Rozanov, V. A. (2015). Rost narusheniy psychicheskogo zdorov'ya v mire—psychiatricheskaya epidemiologiya sovremennosti (Mental health problems growing worldwide—psychiatric epidemiology of novel times). *Uralskiy Zhurnal Psychiatrii, Narkologii i Psychoterapii (Urals Journal of Psychiatry, Narcology and Psychotherapy), 3,* 6–21.

Rozanov, V. A., & Mid'ko, A. A. (2011). Personality patterns of suicide attempters: gender differences in Ukraine. *Spanish Journal of Psychology, 14,* 693–700.

Rozanov, V. A., Valiev, V. V., Zakharov, S. E., Zhuzhulenko, P. N., & Kryvda, G. F. (2012). Suitsidy I suitsidalnie popytki sredi detey i podrostkov v Odesse v 2002-2010 gg (Suicides and suicide attempts among children and adolescents in Odessa in 2002–2010). *Zhurnal psychiatrii i meditsinskoy psychologii (Journal of Psychiatry and Medical Psychology), 1,* 53–61.

Rozanov, V. A., Rakhimkulova, A. V., & Ukhanova, A. I. (2014). Oschuschenie bessmyslennosti suschestvovaniya u podrostkov—sv'yaz s suitsidalnimi proyavleniayami i psihicheskim zdorov'yem ("Life has no meaning" feelings in adolescents—relation to suicidal ideation and attempts and mental health). *Suitsidologia (Suicidology), 5,* 33–40.

Rutz, E. M., & Wasserman, D. (2004). Trends in adolescent suicide mortality in the WHO European Region. *European Child and Adolescent Psychiatry, 13,* 321–331.

Schlebousch, L., Burrows, S., & Vawda, N. (2009). Suicide prevention and religious traditions on the African continent. In D. Wasserman, & C. Wasserman (Eds.), *Oxford textbook on suicidology and suicide prevention. A global perspective* (pp. 63–69). New York: Oxford University Press.

Semenova, N. B. (2013). Kognitivnie factory riska suitsidal'nogo povedeniya u korennyh narodov Severa (Cognitive risk factors for suicidal behavior among indigenous peoples of the North). *Suitsidologia (Suicidology), 1,* 28–32.

Shek, D. T., Lee, B. M., & Chow, J. T. (2005). Trends in adolescent suicide in Hong Kong for the period 1980 to 2003. *Scientific World Journal, 5,* 702–723.

Skinner, R., & McFaull, S. (2012). Suicide among children and adolescents in Canada: trends and sex differences, 1980–2008. *Canadian Medical Association Journal, 184,* 1029–1034.

Slobodskaya, H. R., & Semenova, N. B. (2015). Child and adolescents mental health problems in Tyva republic, Russia, as possible risk factor for high suicide rate. *European Child and Adolescent Psychiatry,* ePub.

Stallard, P. (2016). Suicide rates in children and young people increase. *Lancet, 387,* 1618.

Steel, Z., Marnane, C., Iranpour, C., Chey, T., Jackson, J. W., Patel, V., & Silove, D. (2014). The global prevalence of common mental disorders: a systematic review and meta-analysis 1980–2013. *International Journal of Epidemiology, 43,* 476–493.

Storrie, K., Ahern, K., & Tuckett, A. (2010). A systematic review: students with mental health problems—a growing problem. *International Journal of Nursing Practice, 16,* 1–6.

Tsyrempilov, S. V. (2012). Suitsidogennaya situatsiya v Buryatii: voprosy vliyania etnokulturalnyh factorov i passionarnosti etnosov (Suicides in Republic Buryatya: the impact of ethnocultural factors and passionarity of ethnic groups). *Suitsidologia (Suicidology), 3,* 48–51.

Twenge, J. M. (2011). Generational differences in mental health: are children and adolescents suffering more, or less? *American Journal of Orthopsychiatry, 81,* 469–472.

Twenge, J. M., Gentile, B., DeWall, C. N., Ma, D., Lacefiel, K., & Schurtz, D. R. (2010). Birth cohort increase in psychopathology among young Americans, 1938–2007: a cross temporal meta-analysis of the MMPI. *Clinical Psychology Review, 30,* 145–154.

UNICEF. (2012). *Annual Report 2012 for Kazakhstan,* CEE/CIS. Available from: http://www.unicef.org/about/annualreport/files/Kazakhstan_COAR_2012.pdf

Uzun, I., Karayel, F. A., Akylidiz, E. U., Turan, A. A., Toprak, S., & Arpak, B. B. (2009). Suicide among children and adolescents in a province of Turkey. *Journal of Forensic Science, 54,* 1097–1100.

Värnik, A., Wasserman, D., Dankowicz, M., & Eklund, G. (1998). Marked decrease in suicide among men and women in the former USSR during perestroika. *Acta Psychiatrica Scandinavica, Supplementum, 394,* 13–19.

Vougiouklakis, T., Tsiligianni, C., & Boumba, V. A. (2009). Children, adolescent and young suicide data from Epirus. *Forensic Science, Medicine and Pathology, 5,* 269–273.

Wasserman, D. (Ed.). (2016). *Suicide. An unnecessary death.* New York: Oxford University Press.

Wasserman, D., Cheng, Q., & Jiang, G. X. (2005). Global suicide rates among young people aged 15–19. *World Psychiatry, 4,* 114–120.

Wasserman, D., Rihmer, Z., Rujescu, D., Sarchiapone, M., Sokolowski, M., Titelman, D., Zalsman, G., Zemishlany, Z., & Carli, V. (2012). The European Psychiatric Association (EPA) guidance on suicide treatment and prevention. *European Psychiatry, 27,* 128–141.

Webster, P. C. (2016). Canada's Indigenous suicide crisis. *Lancet, 387*(10037), 2494.

Wiklund, M., Malmgren-Olsson, E. B., Ohman, A., Bergström, E., & Fjellman-Wiklund, A. (2012). Subjective health complaints in older adolescents are related to perceived stress, anxiety and gender—a cross-sectional school study in Northern Sweden. *BMC Public Health, 12,* 993.

de Wilde de, E. J. (2000). Adolescent suicidal behavior: a general population perspective. In K. Hawton, & K. van Heeringen (Eds.), *The international handbook of suicide and attempted suicide* (pp. 249–259). Chichester: John Wiley & Sons.

Wilkinson, R., & Pickett, K. (2009). *The spirit level: Why more equal societies almost always do better.* London: Allen Lane.

Williams, J. M. C., & Pollock, L. R. (2000). The psychology of suicidal behavior. In K. Hawton, & K. van Heeringen (Eds.), *The international handbook of suicide and attempted suicide* (pp. 79–93). Chichester: John Wiley & Sons.

Windfuhr, K., White, D., Hunt, I. M., Shaw, J., Appleby, L., & Kapur, N. (2013). Suicide and accidental deaths in children and adolescents in England and Wales. *Archives of Disease of Childhood, 98,* 945–950.

2

Neurobiology of Stress—From Homeostasis to Allostasis and How Social Environment is Involved

As we could see from the previous chapter, the last 50–60 years in human history are characterized by increase of mental health problems and completed suicides in children and teenagers. Information that about 20–25% of youngsters has some identifiable mental health disorder comes most of all from economically developed Western countries. On the other hand one should keep in mind that objective studies of prevalence of mental health are also more popular in these countries, therefore high prevalence rates may be misleading. Prevalence in the other parts of the world may be even higher, especially when Western culture and global economics capture new regions of the world. Growth of suicides among the teenagers is observed most in the countries which have endured so-called "transition period"—fast and painful transition to market economy, which enhanced social differentiation and inequalities and caused considerable tension in society by strengthening feeling of injustice. Youngsters are sharply reacting to changes in the socioeconomic status of parents, besides the fact that they may feel their own social and property status quite well in different situations—at schools and within their reference group. All this is associated with a variety of feelings and frustrations which may be characterized as psychosocial perceived stress. In this chapter our purpose is to show that the main reason of the observed trends is the psychosocial stress, which is typical for the modern civilized society and is obviously accumulating in the last decade. Though terms "perceived stress" and "psychosocial stress" are widely used in psychological, social, and biological context, this construct still need delineating due to its complexity and versatility.

Stress and Epigenetics in Suicide. http://dx.doi.org/10.1016/B978-0-12-805199-3.00002-6

When speaking about this type of stress we consider it reasonable to discuss it in terms of such phenomenon as modernization. Modernization is understood as process of transformation of traditional societies into more modern and developed ones. Modernization theory supports an idea that traditional societies in this or other form repeat the way or manner modern societies have been developing. Tendencies that we have described, particularly those that compare changes of youth suicides in different parts of the world, existing time shifts, geography and timing of peaks, and decreases of suicide rates seem to fit this model. Ample evidence that young representatives of indigenous ethnics are at highest risk of suicide is another indirect justification of such approach. Our further discussion will go around such problems as the role of stress in suicide, stress-vulnerability, its origin, and relevance for suicidal behavior. We are going to focus on the role of genetic and epigenetic factors in stress-vulnerability and resilience, possible ways that may enhance resilience and their relevance for prevention programs. Our main aim is to integrate biological, psychological, and social data, and to outline ways how cognitive and emotional factors that are intimately associated with all earlier mentioned domains can be involved in such multilayered and multifaceted phenomenon as suicide. In our modeling stress and particularly perceived stress possesses a central role and the whole set of interactions and influences that persist across generations and is understood as the main driving force of negative development and mental health problems and suicides accumulation in the younger generations.

PSYCHOSOCIAL STRESS AND MODERNIZATION

Freud has once stated that civilization makes the person neurotic. Freud's thought can be understood differently, considering basic distinctions between humans and animals and between human society, which existed prior to the beginning of an industrial era and society of a modern type. In the widest evolutionary approach one of the greatest distinctions between animals and humans lies in critically differing level of a stress or in essentially differing type of stress. This question has been discussed by many authors. For instance, R. Sapolsky indicates that the stress in fauna is usually a short-term (though potentially deadly, such as attack of a predator) event, but the overwhelming part of the life-time of animals is devoted to nutrition, reproduction and relaxation. What is most important, as the author of the popular book "Why Zebras Don't Have Ulcers" claims, is that animals after each stressful situation do not ruminate what would happen if they couldn't escape from a predator and they are never anxious in advance, expecting new misfortunes or being emotionally disturbed about future (Sapolsky, 2004).

The human being in this sense is an example of the opposite—the most part of our time we stay in tension, while periods of rehabilitation are getting shorter and shorter. In a shorter historical perspective, since humanity has shifted toward modern types of labor that demands high mental activity, the possibility to escape from overwhelming thoughts and to relax have shortened substantially. The majority of our negative emotional experiences are associated with reflections how to cope with freight of duties and responsibilities. Modern humanity is disturbed by anticipated problems and regrets about unused opportunities or about wrong decisions. Additionally, in modern society not only high intensity, but also high variability of activity is observed. Fast changes of work tasks and new challenges demand constant adaptation, while high level of competitiveness and constant tension together with enormous information flow impose additional requirements (Lundberg, 2006).

Continuing Freud's thought, it is possible to tell that main advantages of western civilization based on liberalism—free access to information, freedom of enterprise, high level of technologies, and especially following the socially approved behavioral models and consumption strategies,—generate high level of stress associated with competitiveness, inequalities, constant tension, high pace of life, information overload, high responsibility for decisions taken, instability, and uncertainty. It gradually turns into a threat to the well-being of the individual. Our body is designed to confront sudden physical threats and endure protracted physical activity; today however, we are increasingly exposed to psychological and mental stressors and are captured by hypodynamia. Civilized humanity faces natural biological and physical stressor less and less, while pressure of psychological factors caused by economic situation, work overload and fatigue, information technologies, and interpersonal conflicts are becoming more and more common. It does not mean of course that we cannot experience typical types of stress, like running for our lives to escape a natural disaster or hostile attack. However, majority of human population, which is currently about 8 billion, experience stress as stressful life events, work overload, loss of control, economic problems or, in the worst variant, poverty, unemployment, conflict, and hopelessness.

On the other hand stress of modern life, being psychosocial by nature, realizes its action by utilizing quite definite and conservative biological mechanisms that are inherent to all mammals. This contradiction, in our opinion, is the main pathological factor of the present times. Current understanding of stress continues ideas of two physiological concepts developed at different times and aimed to explain adaptation of organism to new, unexpected, and generally harmful threats. One is the "fight-flight response"—reaction described by Walter Cannon in the beginning of the 20th century, which is associated with sympathetic autonomous system hyperactivity and blood adrenalin elevation. Another is the so called "general adaptation syndrome"—the more broad concept developed by Hans Seyle a couple of decades later, which have incorporated fight-flight response as an initial stage of every reaction to danger and which focused on cortisol as main stress hormone that starts its activity in case of prolonged action of stressor. The concept of a stress, widely published by Seyle both in scientific and popular forms, and explanations of its role in the development of various diseases has attracted a lot of attention of physiologists, psychologists, and sociologists. Three distinct traditions in the field of stress research have soon emerged—environmental tradition, focusing on environmental events (stressors) that demand adaptation, biological tradition that focused on the essence of physiological reactions in the body in response to stressors, and psychological tradition, which discusses psychosocial stress and individuals' subjective evaluation of the possibility to cope and emotional reactions associated with it. The most widespread modern interpretation of a stress is connected with the cognitivist direction in psychology which proposed modern definition of stress. According to it, stress arises when perceived requirements and demands to an individual exceed the behavioral and emotional resources of the personality. Stress-reaction emerges when the person realizes inability to cope with a problem, to overcome frustration, or to avoid its negative consequences due to the lack of an opportunity to control any event, process, or a state (Lazarus & Folkman, 1984). In other words, situation becomes stressful only as the result of its cognitive appraisal, which includes initial evaluation of the situation as harmful or threatening and subsequent evaluation of possibility to cope. Continuous reevaluation of the situation and coping resources based on new information is referred as reappraisal. In great majority of situations stress is the result of intuitive and even unconscious feeling (or more cognitive understanding) that coping is impossible. Emerging fear, anxiety, and depressive thoughts may add to the existing level of internal feeling of danger becoming the basis of perceived stress, that is, feeling and understanding of how much stress we are experiencing. Perceived stress is measured not by accumulation of stressors (for instance, negative life events), but by summarizing uncontrollability and unpredictability of one's life, one's ability to deal with problems and difficulties. It is based on general perception of stressfulness of one' life and ability to cope and in such form it is represented in the most popular Perceived Stress Scale (Cohen, Kamarck, & Mermelstein, 1983).

This does not exclude the role of common physical, chemical, and biological stressors, which traditionally include noise, vibration, cold or heated climate, air pollution, illness, injury, etc. However, in the modern society, which is technically equipped, many of these stressors are under control. On global scale cognitive and emotional stressors, such as high performance requirements, bundle of problems to be solved, something that requires immediate response and generates high anxiety and tension or fear of failure is much increasingly important. All these types of stressors are usually generated by our living and working conditions, consequently psychosocial stress is often understood as work-related stress. It is considered that work stress occurs when there is a combination of high demands (high output requirements and multiple responsibilities) with the inability to influence or control (low task variety and rigid system of control how the work is done), or when feeling of injustice is dominating—the imbalance between effort spent and reward (Theorell & Karasek, 1996; Siegrist, 1996). On the other hand work situations, though very important, are not the only source of psychosocial stress, it is only one of the most obvious factors that make perceived stress exposed and which contribute to a chronicity of stress. Though psychosocial stress can be understood differently by representatives of different traditions, in general there is a consensus that it is a "human" stress in which the social component is crucial (work, society, relations, self-realization, life goals, and achievements, etc.). For a human being, societies, communities, families, work, relations, and higher level life goals like self-realization form the primary context of existence, therefore stress produced by failures, frustrations, break of relations, and problems in life are largely dictated by these wide social factors.

The concept of psychosocial or perceived stress attracts a lot of attention and is often used to explain the health problems of modern humanity. In this connection it is interesting to analyze the model that link together psychosocial stress, behavior, and health, proposed by Finnish sociologists P. Martikainen, Bartley, and Lahelma (2002). The proposed model (Fig. 2.1) takes into account the biopsychosocial nature of individual, linking psychological processes, and social phenomena with biological mech-

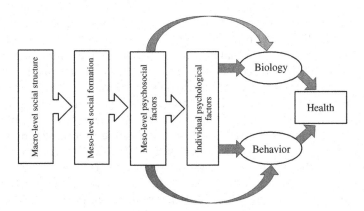

FIGURE 2.1 **Psychosocial pathways leading to health or disease.** *Source: Adapted from Pekka Martikainen et al. (2002).* International Journal of Epidemiology, 31, *1091–1093, by permission.*

anisms and behaviors that are associated with stress. In this model the authors suggest the distinction between macro-, meso-, and micro-levels social factors that influence modern humanity.

Macro-level is determined by the general type of the social system. It includes, in particular, such factors as ownership system, control over resources and businesses, legal and welfare structures, and other mechanisms that determine income distribution and various aspects of property and social relations between groups and individuals. In other words, it is the socio-economic model of society. Meso-level factors are represented by social networks and supportive psychosocial resources. They are partly linked to macro-level factors and include work control with its effort/reward balance, security and autonomy, home control, and work-family conflict. At the individual level psychosocial stress manifests itself as perceived threat to social status, self-esteem, sense of dignity, and similar feelings and emotions. The importance of macro- and meso-level factors, as authors argue, is that they influence the perceptions and psychological processes of the individual. On this lowest level perceived and subjective feelings may "switch on" biological stress mechanisms in the body. These conservative biological mechanisms are supplemented by rather typical behaviors that include enhanced risk-taking, addictive, and self-destructive behaviors, which are moderated by psychological peculiarities of the personality, cognitive style, culture, traditions, etc. Finally, reproduced in more protracted lifestyles they determine health or disease. Summing up, large-scale social effects through meso-level factors, which are represented by psychosocial circumstances, influence individual biological and psychological mechanisms and involve behaviors, for instance, risky behavior, which in turn may determine health or disease in a longer run (Martikainen et al., 2002).

Stress that depends on the social environment cannot be completely pictured if we concentrate only on working environment and do not try to outline some other, sometimes more global factors that have evolved with the beginning of the industrial era. These factors may be understood and characterized within the concept of modernization. Modernization is understood an inevitable process that changes traditional societies into more economically developed and, as proponents of this theory argue, progressive societies. Modernization theory links together economic, cultural, and political changes, in the historical perspective modernization, postmodernization, and second modernization is postulated (Inglehart, 1997). If the initial stages of modernization were associated with transition from agricultural to industrial societies, further waves are understood as transition to information and knowledge society. Supporters of modernization theory consider that modernized states are wealthier and more powerful and that their citizens are enjoying higher standards of living and freedoms. But together with more liberal freedoms, technological and

TABLE 2.1 Some Most Evident Differences Associated With Modernization

Traditional societies	Modernistic societies
• Big and stable families • Adherence to traditions and strict rules, morals, religiosity • Stability, slow pace of social changes • High family and community cohesion • Low anxiety and depression	• Unstable and incomplete families • Competitiveness, priority of personal success, liberalism • Mobility, high speed of social changes • Individualism, materialism • High anxiety and depression • Technologies and information pressure

economic development, modernization is associated with several other factors that enhance psychosocial stress. We have tried to outline some differences between traditional and modernistic societies that are relevant for our analysis here (Table 2.1).

In view of changes that have happened for recent century or two, a period that covers all waves of modernization, we can notice several factors that may mean even higher level of psychosocial stress. A century or two is a very short time in the history of the humanity and civilization, but a very reasonable period of time from the point of view of possible accumulation of epigenetic transformation with transgenerational inheritance (which will be extensively discussed in further chapters). Three global factors can be pointed as following: (1) the decline in the level of social support; (2) change of the nature of labor; and (3) global chronicity of stress. The first factor is due to the so-called demographic modernization—transition from high mortality, high birth rate and low life expectancy to low mortality, low birth rate and high life expectancy. Life expectancy is growing almost linearly for the last century with a concomitant decrease of the number of children per family. As a result the number of people surrounding each individual has diminished dramatically, which means that there are less close relatives who can provide support and sympathy for each personality (Kirchengast & Haslinger, 2010). This is happening on the background of urbanization and general population growth. By now, globally, more that 50% of people are already living in the cities, while in some regions of the world urbanization reaches 80%. It results in unprecedented misbalance between overcrowding and loneliness. Staying "alone in the crowd" has reached its highest level, humanity have never lived like that in its whole history. This creates new reality, which is a challenge for well-being of big contingents.

The second factor relates to the fact that if even 100–150 years ago the vast majority of people in the world have been involved in very traditional (often agricultural) types of labor, which implied a lot of physical activity, while now the majority of people are working "in the office." This dramatically increases the burden on the neural network structures and nervous system in general, resulting in the growth of diseases, which are often characterized as "diseases associated with the brain" (World Health Organization, 2008). Under this nontypical term many existing conditions are united, both neurological and psychiatric, from addictions to different dementias. Such global transition to the intellectual types of work has caused inevitable changes in the education system for younger generations, which dramatically increased the information pressure on the children and adolescents in schools and on young people in the universities. This contributes to higher tension and pressure on young people, which is depicted in numerous reports, how modern youth is overwhelmed by school tasks and the level of perceived stress in the young people mentioned in a previous chapter. On the other hand, it leads to decreased physical activity, which gives the way to sedentary lifestyle, thus weakening its protective role on dementias, stress disorders, and depression. All this is happening against the background of the third factor—constantly growing level of global psychosocial stress, induced by social and economic circumstances. The reasons for this are diverse and interconnected with other mentioned factors.

Modernization is closely associated with economic development and higher standards of living. But in a modern world, it also results in deeper differentiation between rich and poor and sharpens inequalities between and within the countries, both developed and developing. Moreover, higher living standards, better nutrition, and better education did not lead to higher birth rates in economically rich regions. On the contrary, percentage of "child free" families is growing together with higher general life expectancy. As a result, we are facing the ageing societies with accumulating problems due to diminishing percentage of economically active population. Lower number of close relatives surrounding one person in the developed world means not only lower social support, but also fewer contacts between young and old, growing gap between generations, which as many suicidologists consider, can contribute to suicides. High birth rate and high mortality still can be seen in poor countries and regions of the world together with low life expectancy, but here another bundle of problems is accumulating—total poverty, poor nutrition, clear water shortage, low level of medical aid, and general instability, which also means high stress though of different nature.

Demographic modernization is only one of manifestations of those fast and versatile processes which take place in the modern world and which affect both youth and elderly. Russian physicist Sergey Kapitsa have pointed on another distinctive feature of the present. It is a "compression of historical time"—an intensification of all kinds of activity and the accelerated introduction of various technological innovations (Kapitsa, 2004). S. Kapitsa pointed that it took centuries to introduce technological innovations in the early period of civilization, some 150–200 years ago, with the beginning of the industrial era important innovations were introduced during the life of 1–2 generations, while in recent decades during the life of one generation such great number of the novelties completely changing our life are introduced, that not all are capable to adapt to it. Other signs of time compression are dynamism in labor market and constant need to be trained and retrained, globalization and loss of identity. As Kapitsa argues, the "...extremely fast pace of changes have led to crisis and stress at the individual, on the family and society level, which is expressed both in infantilism and in loss of "reference points," and in many manifestations of public consciousness, as well as in art and literature ..." (Kapitsa, 2004). Loss of "reference points" can be associated with the fact that at high pace of development of society economics pragmatic values start to prevail, while ideologies, morals, and moral principles are eroded and becoming blurred. It may be said that the main semantic markers of society are losing their significance, become soft and indistinct. These deep believes, concepts, and values are formed and consolidated in the societies during historically prolonged periods and are supported by deep traditions, life practice, and conservatism, while under conditions of the increasing rate of modernization, there is simply not enough time even for understanding the nature and essence of the changes that happen and for development of attitudes toward them. As a result, new generation is losing life reference points, change from intrinsic to extrinsic values and goals, experience loss of meaning, and long-lasting purposes in life. These ideas are close to those expressed by A. Toffler in his famous "Future Shock."

At last, one more sign of modernization is exclusively powerful influence of information technologies. This is a factor that strongly enhances exposure of very wide public and especially young people to many stress-inducing factors, like inequalities, propaganda of consumerism, or other examples from "adult life," including sexual behaviors, risk-taking behaviors, and self-destructive behaviors. Another factor that is typical from modern mass media is that they suggest a stream of short clips of reality that create fragmentary mosaic of events and happenings. As many authors argue, this may impair awareness of causal relationships between events and have an impact on the type of thinking, creating so called "clip thinking." The last is characterized by impairment of conceptual understanding of the world in all its' complexity, when logic or internal associations are substituted by fragmentary representations. This problem recently is acknowledged mostly by pedagogical community, though it may be relevant in a wider context (Semenovskikh, 2014).

All the earlier mentioned factors influence everyone today , exerts impact on the big contingents and impact different sides of human consciousness, from logic to emotions and aspirations. For instance, modern economic model demands growing consuming, and to meet the needs of companies mass media and the Internet through advertising and spectacular programs promote consumerism, hedonism, and general aspiration to comfort. Modern civilization turns to be a civilization of comfort, which leaves less space for "big meanings." As a result, traditional values of self-restriction and regulation of one's hedonistic impulses are undermined and appear outdated. No surprise that it has negative consequences in many spheres and inevitably promote frustration. Ubiquitous Internet content makes a substantial contribution to these processes by making this information flow easily accessible to youth in all parts of the world, including developing countries and even in the most traditional ethnics, occupying distant areas, replacing meaning and significance with entertainment, real life with virtual reality. With regards to the problem of suicide it is quite appropriate to remind, that Korean authors consider suicide exposure by mass media one of the leading factors of negative tendencies with suicides in young people (Park, Im, & Ratcliff, 2014). Quite recent data from USA confirm that suicide exposed individuals are twice more likely to have diagnosable depression and anxiety compared to suicide unexposed (Cerel et al., 2016). So what can be said about youngsters in some distant region of Siberia or Venezuela who just received their first IPhone with Mediametrics software, which posts suicide cases constantly.

These and other factors of modernity determine increased level of psychosocial perceived stress. In wider terms it is possible to call it a modernization stress. While being psychosocial by nature, this stress is characterized by involvement of conservative biological mechanisms and leads to deterioration of physical and mental health of the big contingents of people. It is therefore important to look at the concept of stress from this point of view before we start discussing the role of stress and epigenetics in suicide specifically.

NEUROENDOCRINOLOGY OF STRESS RESPONSE— HOW PSYCHOSOCIAL STRESS AFFECTS HEALTH

After original formulation of the concept of stress by Hans Selye as of nonspecific reaction of the organisms to harmful influences, dozens of different approaches to conceptualizing stress have been developed by physiologists, sociologists, and psychologists. There are inevitable difficulties in understanding stress due to complexity and interrelations between the reason and the consequences of this reaction, variety of harmful factors and often not uniform internal reactions to them. One of the important step was differentiation between stressor (influencing factor, presumably external) and stress-reaction (internal mechanism that is developed in response to stressor). On the other hand some stressors may be internal and may belong to stress-reaction mechanisms or may be tightly associated with them. It includes emotional reactions and cognitive mechanisms that moderate, modulate, and often mediate influences of stressors, which are biological and psychological at the same moment.

Nevertheless there are two main definitions that are prevailing. From the physiological point of view, stress is often defined as a state of threatened homeostasis of the organism which is the result of necessity to adapt to extrinsic or intrinsic, real, or perceived stressors (Charmandari, Tsigos, & Chrousos, 2005). Psychological approaches focus on cognitive component; therefore stress is defined as a mismatch or imbalance between the demands that are placed on the individual and ability to cope with the problem (Danielsson et al., 2012). It is important to mention, that this ability is often an estimation, a perceived ability, which is dependent on the previous experience, memories, and strategies one have used to solve problems and to overcome previous obstacles and problem situations. Within this context two aspects of stress-reactivity are important: stress-vulnerability and stress-resilience. As recent studies have proved, both are dependent on the previous experience and are largely determined by early life conditions (Sapolsky, 2015; King, 2016).

Mechanisms, that underlie individual vulnerability and resilience, may include biological, psychological, and social component. Many studies recently provide evidence that enhancement or attenuation of stress reactivity is associated with epigenetic effects, the subject that will be discussed in detail in further chapters. Meanwhile we will concentrate on physiological mechanisms of stress response and on pathogenetic pathways, which lead to different stress-related diseases, including mental health problems and disorders. It is important to keep in mind that stress-induced physical and mental diseases or stress-related states and conditions, like burnout syndrome or chronic fatigue, subthreshold mental health problems or clinically distinct mental disorders may play the role as risk factors of suicide.

Each stressful situation leads to activation of biological systems of stress response in the organism which have urgent and more protracted components. These systems were developed evolutionary as the main mechanism of survival when exposed to short-time threats, like predators' attack or natural disaster, and in a more long run for being prepared and vigilant in a hostile environment. The type of stress we are exposed today is mostly a prolonged and protracted unavoidable stress; therefore its underlying biological mechanisms cannot be switched off immediately. On the other hand modern human stress consists of a continuity of more or less acute reactions, for instance, in response to stressful information messages or frustrating thoughts and feelings. Recent decades of studies of stress neuroendocrinology elucidated main mechanisms of stress reaction and added a lot to initial physiological concept developed by Hans Selye.

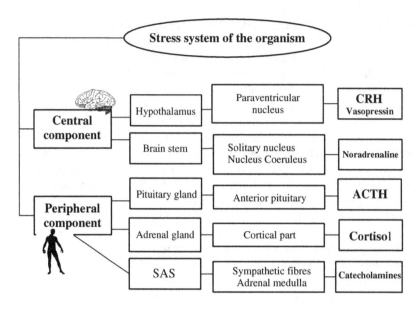

FIGURE 2.2   Central and peripheral components of stress system of the organism.

Stress systems of the organism consist of the central and peripheral components. The central component is located in the hypothalamus and brain stem—phylogenetically rather old parts of the brain. It includes several types of neurosecretory cells in paraventricular nucleus and lateral parts of hypothalamus that produce corticotropin-releasing hormone (CRH) and vasopressin, and noradrenergic nuclei in the medulla and pons of the brain (solitary nucleus, locus coeruleus), that produce noradrenaline. Locus coeruleus also contains CRH-producing cells (Fig. 2.2). Discovery of neurosecretory cells and their role in stress response made it clear that brain is actually an endocrine organ and helped to develop complete pathophysiological scheme of stress-reaction (Sapolsky, 2015).

The peripheral components of stress response system include pituitary gland, which produces ACTH, and adrenal gland, particularly its cortical part, which produces main effector hormone cortisol (in case of rodents and other nonhuman animals—corticosterone). All together it forms the so called hypothalamic-pituitary-adrenal axis (HPA)—a system of interrelated neuroendocrine signals and feedback mechanisms. Peripheral part also includes efferent sympathetic-adrenomedullary system and components of parasympathetic system (Charmandari et al., 2005). Physiologically stress response system involves components of limbic system of the brain (hypothalamus, hippocampus, amygdala) which has projections both to old parts of the brain (brain stem) and higher centers (neocortex) and is connected with sensory systems. All components of the central part

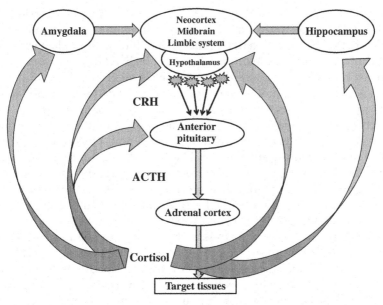

FIGURE 2.3 Hypothalamic-pituitary-adrenal axis negative feedback regulation pathways.

of the stress system are integrated by complex regulatory and autogerulatory interaction, which are still extensively studied. Most important for understanding stress-reactivity is the vertical regulatory mechanism—from central to peripheral parts, from hypothalamus to adrenal gland that keeps the balance and dynamic state of the HPA.

One of the main features of HPA is a mechanism of negative feedback which is based on inhibitory effect of cortisol on CRH and ACTH secretion (Fig. 2.3). Main targets of this regulation loop are extrahypothalamic centers like hippocampus and amygdala as well as hypothalamus and anterior hypophysis itself (Tsigos & Chrousos, 2002; Charmandari et al., 2005). All these structures are especially abundant in cortisol receptors. This feedback mechanism provides for regulation of basal level and for restriction of excessive activity of HPA, thus minimizing metabolic and physiological effects of elevated HPA components. It also enables a dynamic mechanism of multiple pulses of activity of HPA during the day with general diurnal rhythm (early elevation, high level at the day time, and lowering during the night sleep). It is also important in stress situations so far as excessive CRH may influence brain structures producing effects on eating behavior, reproduction, growth, and immune system functions that are beyond its regulatory role in HPA (Tsigos & Chrousos, 2002).

In any acute stressful situation activity of the hypothalamic-pituitary-adrenal axis (HPA) is sharply enhanced. Activation of CRH secretion, initiated by signals from upper centers, leads to immediate release of ACTH in the blood and slightly delayed in time final activation of secretion of cortisol by the adrenals (Aguilera, Kiss, Liu, & Kamitakahara, 2007). Effects of cortisol as of pleiotropic agent are realized due to its interaction with the corresponding receptors extensively distributed in various organs and tissues. Hormone-receptor complexes of cortisol play the role of transcriptional factors that are able to change the profile of activity of the big sets of genes. Due to this and also due to urgent effect of sympathetic innervation component, all main metabolic and physiological stress-related effects of stress are realized (Hayashi, Wada, Ito, & Adcock, 2004; Kumar & Thompson, 2005). These reactions are important for adaptation to living in violent and hostile surroundings, as well as for the action of pathogens, insufficient nutrition, harsh climate, and other factors by adjusting metabolism, emotional, and behavioral responses.

For instance, activation of stress system associated with urgent effects of sympathetic nerves leads to efficient mobilization of metabolically active phosphorylated sucrose caused by adrenal medullar epinephrine. Further stimulation of gluconeogenesis, proteolysis, and lipolysis provide elevation of blood sucrose and fatty acids in a more protracted manner. It is vital for nerve and cardiac tissue nutrition and active functioning and ensures energy metabolism boost on the cellular level. These cellular changes are accompanied by physiological reactions like blood pressure elevation, blood clotting system activation, centralization of circulation (often noticed as paleness of the integuments), activation of immune response, and simultaneous emotional reactions of fear or anger, aggressiveness, and hostility. At such moments many other physiological reactions also become evident—sensitivity to pain is reduced, emotional memory, and attention (concentration) is enhanced to the level of "tunnel vision," muscles of the face and neck are contracted. Therefore, the organism gets prepared for action, possible fight, and gets ready to cope with trauma or injury, blood loss, etc. All energy resources of the organism are mobilized, which cannot last for a long time, so when the stress situation is over, all these functions should be quickly shut down to preserve systems of the organism from exhaustion. Sometimes if the danger is too strong, and coping resources are too scarce, the organism may choose "shutting down" behavior with such symptoms as fainting, dizziness, tiredness, and muscular debilitation (Danielsson et al., 2012).

In cases of protracted or chronic stress constant mobilization may lead to stable domination of degenerative processes over regenerative. All above mentioned energy consuming processes gradually start to "rob" other energy dependent tasks, like body growing and cells proliferation, healing, rehabilitation, reproduction and restoration. Reactions that are beneficial as a short term adaptations, in a long run or in the situation of chronicity are exerting obvious health risks, promoting hyperglycemia, hyperlipidemia, hypertension, and contrinsular effects (Sapolsky, 2004; Meaney, Szyf, & Seckl, 2007). Quite expectedly chronic tension can lead to many diseases which are so typical for modern humanity and are known as "civilization diseases." This is confirmed by association between stressful life events accumulation (objective sign of stressful experiences, which nevertheless do not reflect the complexity of the internal perceived stress and coping abilities) and such conditions as diabetes, cardiovascular diseases, atherosclerosis, periodontal disease, obesity, etc. (Tosevski & Milovancevic, 2006). Moreover, chronic stress is leading to several "new" syndromes that cannot be identified as classical diseases, but which are very prevalent in modern populations, like neck and back pain, chronic fatigue, sleep disorders, emotional disturbances, as well as mental health problems.

When speaking about the role of stress as a pathogenetic mechanism of human diseases it is necessary to encounter that outcomes depend essentially on the personality characteristics. For example, cardiovascular problems, such as hypertension and myocardial infarction are closely associated with "type A" personality (high level of emotional reactivity, aggressiveness, competitiveness, aspiration to achievements, etc.). A number of objective studies show that such individuals have almost twice higher risk of a heart attack while having such risk factors as metabolic syndrome, sedentary lifestyle, smoking or alcohol consumption. Such traits as irritability and hostility appeared to be the most dangerous personality variables that enhance health risks (Lundberg, 2006). A number of studies shows, that the work associated with significant stress (a combination of high demands and low possibility to influence the situation, or an imbalance between the level of effort and reward) is associated with a higher risk of heart attack, and this are especially relevant for males (Siegrist, 1996; Belkic, Landsbergis, Schnall, & Baker, 2004). Thus, psychosocial stress is a direct risk factor for heart problems. The same kind of stress (evaluated as accumulation of stressful life events), and belonging to the "type A" personality is strongly associated with the risk of cerebrovascular diseases, and in contrast to a heart attack, in this case risk does not depend on sex (Belkic et al., 2004).

Other studies have shown that psychosocial stress enhances the risk of viral infections, from the banal cold to HIV/AIDS (Antoni, 2003; Cohen, 2005). This is understood as an immunosuppressive effect of stress—a well-known effect of corticosteroids, especially when they are used as a medication. It contradicts results of other observations, which point on activation of inflammation is chronic stress and which consider autoimmune and neuroinflammation processes as main signs of modern psychosocial stress. This contradiction may reflect possible differences in effects of stress hormones in different types of stress due to dosages, chronicity or locality of the action. Many other pathological conditions, diseases and maladaptive biological manifestations are also explained by elevated level of stress in the modern society. In the recent decades human reproductive health is seriously deteriorated, which undermines prospects of homo sapiens as a species. Substantial part of contemporary problems in reproductive health, like female infertility and decreased sperm quality in men, the severity of menopausal symptoms and impaired fertility in men with age are also strongly associated with various forms of stress (Podolska & Bidzan, 2011; Gollenberg et al., 2010; Binfa et al., 2004; Hall & Burt, 2012). Psychosocial stress plays an important role in the pathogenesis of irritable bowel syndrome and periodontal disease, the prevalence of which in modern populations is growing on all continents (Surdea-Blaga, Băban, & Dumitrascu, 2012; Goyal, Jajoo, Nagappa, & Rao, 2011). Reactivity of HPA and psychosocial comorbidities are also reported to be strong predictors of skin diseases, particularly psoriasis (Schwartz, Evers, Bundy, & Kimball, 2016).

Finally, more and more empirical data accumulates to support the hypothesis about the role of stress in the genesis of various forms of cancer and temp of aging. Though results of epidemiological studies are inconclusive, there are indications of the association between stress and cancer, which may be partly explained by changes in lifestyle and strategies of coping with stress (Johansen, 2012). Metaanalysis of 165 studies have revealed that stress-related psychosocial factors are associated with higher cancer incidence in initially healthy populations, poorer survival in patients with diagnosed cancer, and higher cancer mortality (Chida, Hamer, Wardle, & Steptoe, 2008). As to aging, several recent studies focused on telomeres length in leucocytes as the sign of early aging. These conservative nucleotide sequences forming endings of chromosomes are known to be shortened after each cell division thus carrying out functions of the cell "clock" or memory of the number of previous divisions. Since the discovery of this phenomenon several studies have revealed close relation of telomere shortening to common psychosocial factors. Accelerated telomeres attrition was found in higher work-related stress, in burn-out syndrome (Ahola et al., 2012) and in general psychosocial stress and dysfunctional lifestyles in longitudinal studies (Revesz, Milaneschi, Terpstra, & Penninx, 2016).

This short review gives an impression about the nature of links between psychosocial stress and most prevalent diseases in the modern society. Bodily responses to stress appear to be those internal factors that in the long run trigger mechanisms that, instead of adaptation, which is the ultimate goal of stress-reaction, cause maladaptive development thus producing multiple adverse effects. On the other hand, modern stress is an interactional process that is described as an imbalance between the environmental demands and the individual's perceived resources to cope with these demands (Lundberg, 2006). It shifts focus to the area of mental manifestations, to perceived, psychic or mental stress, that exists literally more "in the head" than in the environment. It is associated with perception of poverty, exclusion, alienation, inequalities, frustration, injustice or loneliness, and abandonment, or maybe the subjective feeling of stress that makes life so stressful.

When speaking about this side of stress, two big questions rise. First is regarding interaction of such stress with biological mechanisms mentioned earlier, that is, to what extent perceived stress trigger internal stress reaction. Second is regarding how brain itself is involved in development of this vicious cycle. Brain is the main organ that regulates stress reaction and, as we can see from modern neuroendocrinological data, is the main target of stress. Stress was evolutionary designed as a source of adaptation, but in modern life turns into one of the reasons of diseases, producing maladaptive outcomes. It is therefore important to look at the biological mechanisms of stress from this point of view. Recent studies have provided important explanations which are based on understanding how homeostasis gives way to allostasis and how neuroplasticity, life experiences, and epigenetics are involved in development of mental stress. This will be discussed further.

FROM HOMEOSTASIS TO ALLOSTASIS AND HOW SOCIAL ENVIRONMENT IS INVOLVED

The concept of homeostasis focuses on the idea of maintaining stable internal environment of the organism by timely involving biological mechanisms that counterbalance evolving changes and perturbations caused by stressors. Among these stressors may be physical (noise, heat, vibration, temperature), chemical (environmental pollutants, toxic substances, medications overdosed), biological (exhaustion, diseases, both infectious and noncommunicable), cognitive (problems to solve, mental demands, high commitments, and responsibility), emotional (anxiety, fear, frustration) and psychosocial or perceived (time and information pressure, interpersonal conflict, uncertainty, alienation, lack of control, inability to cope with problems, overcrowding or loneliness). In each case short term stress, when passed successfully, usually leads to positive health results by creating memories and experiences of successful coping, while chronic and constant stress in a long run leads to disease. It is important that homeostasis may be maintained in both cases though internal mechanisms that keep it in certain limits may be rather differing. To explain how stress responses gradually are shifting from health-promoting to health-damaging, a concept of allostasis and allostatic load is used (McEwen, 1993). Allostasis was initially understood as operating range, or the ability of the organism to increase or decrease vital functions in response to a new state or challenge. Operating range is higher in health and lower in a disease, and is larger in younger than in older individuals (McEwen, 1993).

The concept of allostasis appeared to be very relevant for understanding health-damaging effect particularly in case of chronic psychosocial stress, when social environment is the main source of stressful experiences (McEwen, 2012). As McEwen conceptualizes, long term effects of unavoidable and repeated stress leads to "wear and tear" of the biological systems of the organism which is followed by physiological, emotional, cognitive, and behavioral consequences. This wear and tear is the results of complex interactions between different parts of regulatory neuroendocrine system of the body. The essence of pathological development is that each new steady operating state induced by stressful challenges and demands leads to higher level of the activity of counterbalancing systems of the organism. As a result in progression of stress balance between systems is achieved on a substantially higher level of their activity. An analogy to this is a see-saw, which is balanced by two heavy weights as compared with the same see-saw balanced with much lower weights. The strain in the body due to constant ups and downs and huge efforts of regulatory systems to keep the balance is defined by the author of the concept as "allostatic load" (McEwen, 1993). The most serious negative results may follow if one of the counterbalancing systems is compromised or suddenly break.

Speaking about allostatic load McEwen quite reasonably takes as an example stress associated with a typical busy workweek. Cyclic type of occupational stress with high demand, low control and constant pressure is a known risk factor for coronary heart problems and cerebrovascular diseases, associated with atherosclerosis (Kamarck, Muldoon, Shiffman, Sutton-Tyrell, & Gwaltney, 2004). Actually working week load is thought to be quite well balanced by week-ends, and it is supported by all previous practice. But in fact the problem is deeper; it seems to be imprinted in the biological mechanisms that are confronted by new type of stress that is inherent to modern life. When one heavy weight is removed from the see-saw (relief and lowering of stress exposure) organism needs time to adjust for new balance, and in this situation different problems with seemingly unconnected systems may evolve. For instance, it is well known that after serious strain long waited resolving of the situation and obvious relief is often followed by unexpected health problem, like infectious disease or depression. Really, colds and other infections in modern life often manifest themselves on week-ends or on vacations, after prolonged period of intense demand, while depression can become most evident shortly after the holiday, etc. (McEwen, 1993).

It is often not easy to find physiological explanations and confirmations for these very logical and supported by practice theoretical concepts because of the variability of stressors, different timing of the events and individual psychological features and cognitive styles which moderate or modulate effects. Behavioral style is an important variable as it was previously specified in scheme of health consequences of psychosocial stress (Fig. 2.1). For instance, smoking and alcohol consumption, which is a common stress-limiting behavior, may be partly adaptive for a while and even may lower the risk of heart attack, but may contribute to higher risk in a long run. This may be modified by cognitive styles, which may promote more severe risky behavior and contribute to earlier disease, or on the contrary, due to turning to healthy behavior, lower risk or remove terms of the onset of the disease to older age, and even eliminate the risk. However, there are many known physiological facts that may be well interpreted in terms of the theory of allostatic load. Abnormalities in the immune system, in the cardiovascular system, and in muscle and adipose tissues, as well as chronic pain (often difficult to explain so far as it has no physical basis at the initial stage) that are inherent to modern human populations may be caused by complex interactions between mediators and effectors of the stress response of the organism. For instance, elevated cortisol secretion promotes insulin secretion, combined action of these factors accelerate atherosclerosis, which in a long run may lead to heart attacks. Frequent elevation of insulin stimulate contra-insular factors like glucagon, while sympathetic impulses by themselves act against insulin, and in a long run it leads to type 2 diabetes, especially if overeating induced by anxiety is involved. Glucocorticoids and catecholamines are synergic toward some immune system functions, but antagonistic regarding others, producing both antiinflammatory and proinflammatory effects in different situations. Parasympathetic innervations counterbalance adrenergic effects and have antiinflammatory action. Low levels of cytokines are neuroprotective while their overproduction may cause neuroinflammation and, as new data suggest, depression (McEwen, 2012). High stress load impairs the functioning of muscle spindles which leads to chronic muscle stiffness and enhanced firing in pain nerves, thus producing higher sensitivity to pain (Danielsson et al., 2012). Muscle stiffness and failed attempts to relax (or behavioral and cognitive inability to relax) in a long run impairs joints functions. After joints start to be compromised structurally, their degradation may add to autoimmune disturbances enhancing pressure on the immune system and promoting systemic inflammation and further bone tissues degradation. This is only part of possible interactions of counterbalancing systems and their molecular mediators. In every individual case genetic predispositions make these interactions and balances rather specific, which produce huge variability of health outcomes and hampers conclusive explanations and models of health and disease outcomes in psychosocial stress.

Except genetic makeup of the individual all these effects are strongly moderated by social context, for instance social status, early life context and learned social history. Social position and status (dominant or submissive) together with life history and coping skills may act as important moderators in the processing of stress stimuli, which in turn influences the whole biological repertoire of stress response. It is supported by studies, which show that stress-related reasons for multiple metabolic disorders like hyperlipidemia, decreased tolerance to insulin, and obesity in human populations are adequately modeled by social stratification (dominant and subordinate individuals) in the communities of apes and rodents (Tamashiro, Sakai, Shiveley, Karatsoreos, & Reagan, 2011). Status syndrome, fight for higher status, and frustration in case of social defeat are considered by Wilkinson and Pickett (2010) as main moderator of association of somatic manifestations and mental disorders with social factors (social inequality, low socio-economic status, work load, economic difficulties, level of social support and so on). Subjective component, like sense of well-being, chronic fatigue, and depressive symptoms, are associated with severity of symptoms of metabolic syndrome, diabetes, dyslipidemia, and coronary heart disease in population-based studies (Watanabe, Stewart, Jenkins, Bhugra, & Furukawa, 2008; Wiltink et al., 2011). Summing up it may be said that the model of psychosocial stress is the best to describe the role of such factors as social status and related discrimination, stigmatization, and prejudices on probability of different diseases and conditions (Schwartz & Meyer, 2010). Mental health problems among them are the most important.

Besides physical health impairments, frequent consequences of stress are emotional, behavioral, and cognitive disturbances, which are very prevalent in the modern society, though they may not even lead to diagnosed diseases. Many studies link together psychosocial stress and mental health problems like anxiety, depressive symptoms, and addictions with working environment and labor relations, unemployment and social insecurity, social cohesion and social exclusion, role of the transport problem in cities and other urbanity factors. All these factors, as shown by objective research, not only contribute significantly to morbidity and mortality from cardiovascular diseases, cerebrovascular diseases and diabetes, but have even more significant impact on the mental health of populations. Mental health problems are considered the most common and universal effects of psychosocial stress, especially when it comes about chronic, constant influence of stress factors. They include burn-out syndrome, chronic fatigue, or sleep disorders, as well as more severe manifestations. Evolving of these conditions are well described by the stress process model, which incorporates multiple levels of support and stress at the individual, family, and community level, with a focus on predicting mental health outcomes (Katerndahl & Parchman, 2002; Anderson, 2004). This model assumes that stressors and coping resources, which are dependent on social context, are especially important for mental health risks. Our further discussion will focus mostly on the intrinsic mechanisms of the most prevailing stress-induced mental health problems, like depression, anxiety, addictions, and suicidality.

STRESS AND THE BRAIN—MENTAL HEALTH CONSEQUENCES

One of the most common signs that stress-related state of the personality is getting worse, are growing subjective complains—poorer performance, chronic fatigue, disinterest, dejection, lack of work motivation, sleeping problems, pain, sometimes dizziness and chest pressure, often accompanied with anxiety about one's health (Lundberg, 2006; Danielsson et al., 2012). All these signs may eventually transform into depression, burn-out syndrome, psychosomatic, or anxiety disorder. Mental health consequences of psychosocial stress can be understood within the context of HPA regulation and effects of cortisol and other components of HPA, including CRH. Brain is the main organ that organizes stress response by regulating all neuroendocrine body mechanisms involved in the stress-reactivity. Main feedback mechanism—a loop of reverse action of cortisol—also ends in brain structures. Besides direct and rapid inhibition of CRH production by hypothalamic paraventricular nucleus neurons via its receptors, cortisol also involves several structures of the limbic system (especially hippocampus and amygdala) and prefrontal cortex where it triggers receptors widely represented in neuronal networks (Tsigos & Chrousos, 2002; Dedovic, Duchesne, Andrews, Engert, & Pruessner, 2009; Herman et al., 2016).

How this regulation mechanisms influence brain structures and what will be the functional outcomes of this loop depends on several factors and is closely related to the phenomenon of neuroplasticity. Revolutionizing studies that have proved that brain is not a static but a constantly changing organ, where new neurons and synapses are born in structures that are responsible for experience accumulation, provided new insights into neurobiology of stress. These processes are associated with formation and remodeling of synaptic junctions, finding by neurons their new counterparts, establishing new synaptic contacts and pruning unnecessary ones (Michmizos, Koutsouraki, Asprodini, & Baloyannis, 2011). Synaptic plasticity is the central mechanism of brain remodeling in response to different stressors. For instance, if HPA response is maintained within adequate limits and cortisol level quickly falls after each stress response, cortisol due to high density of its receptors and effective coupling with Ca-dependent potentiation mechanism in hippocampal dental gyrus promotes formation of emotional memory traces that are probably involved in development of copying strategies (Herman et al., 2016). On the cellular level transient stress promotes neurogenesis by influencing neural stem cells and enhances synaptic neuroplasticity as if brain structures are preparing themselves for new transient stresses. On the contrary, chronic stress prevents neurogenesis and hampers neuroplasticity by suppressing new interconnections between neurons or by pruning existing ones (King, 2016).

Another differentiation may be seen if type of stressor is taken into account. Something that is typical for reactive physical stress may be not the case when organism is challenged by chronic anticipatory or mental (psychological, perceived) stress. Reactive stressors are increasing the demand on the system through real sensory stimuli, such as pain, bodily injury, or an immune challenge (infection), while anticipatory or psychological stressors are associated with social challenges, unfamiliar situations or perceived threats. Existing data suggest that physical stressors predominantly engage phylogenetically old parts of the brain like brain stem and hypothalamus, while perceived stressors mostly involve amygdala, hippocampus and prefrontal cortex—structures, which bear conservative system of reactivity to social disadvantages and threats (Dedovic et al., 2009). If translated into human behavioral and emotional reactions, short term physical stressors (if successfully passed) give a feeling of well-being, effectiveness, and high self-esteem, while chronic perceived stressors induce fear, anxiety, social defeat behavior, low self-esteem, impaired autobiographic memory, ineffective coping, and disturbed cognitive abilities, like impaired planning and decision making (McEwen, 2016).

Results of animal studies of the role of chronic stress in brain and observations in humans using neuroimaging techniques often do not fully coincide, and neuroplasticity in different brain structures is not the same. Great number of studies found that chronic stress associated with psychological trauma, depression, or PTSD leads to diminishing of the volume of hippocampus in humans (Woon, Sood, & Hedges, 2010). In animal studies high concentrations of corticosteroids are found to be destroying neurons in hippocampus thus demonstrating neurotoxic action. It appears that the same tandem—excitatory amino acids and glucocorticoids, which is involved in formation of emotional memory in case of successful coping, under different conditions can produce effects that resemble those registered in traumatic brain injury or hypoxic brain damage (Sapolsky, Krey, & McEwen, 1986). Most obvious neuron loss is found in hippocampus abundant with glucocorticoid receptors (McEwen, 1993, 1999). Shrinkage of the human hippocampus quite expectedly is most often found in mild cognitive impairment, dementias, prolonged depression, and PTSD. On the other hand several stress-related conditions in humans like chronic inflammation, sedentary lifestyle, lack of physical activity, and jet lag may be also associated with lower hippocampus volume (McEwen, 2012). Though hippocampus is the main target of adverse effects of stress hormones in brain, it is not the only one. Decreased synaptic plasticity, dendrites atrophy and loss of total volume of gray matter in stress can be also found in a prefrontal cortex. Such alterations have been found in chronic stress, circadian disruptions, and sleep disturbances and were associated with impairment of cognitive flexibility, fatigue, and decision making problems—all signs of shift work-related impairment (McEwen, 2012). While hippocampus and frontal lobes neuroplasticity is mostly impaired in chronic stress, in amygdala an opposite picture is often found—enhanced synaptic and dendritic processes (Sapolsky, 2015). Thus, chronic stress, which impairs dentate gyrus neurogenesis, causes expansion of dendrites in the basolateral amygdala (Vyas, Mitra, Shankaranayana, & Chattarji, 2002). If translated into human behavior and emotions this may mean feelings of fear, anxiety, and being stressed, enhanced reactivity to angry and sad faces, depressive symptoms, and aggressive bouts. It is also interesting to mention that amygdala is thought to be involved in perception of social situations and composition of emotions associated with social domination and social defeat or submission, all dependent on social judgment (Adolphs, Tranel, & Damasio, 1998). Main targets of stress hormones in brain and emotional-behavioral consequences of their participation in stress-reaction are presented in Fig. 2.4.

Varying picture of effects of cortisol in stressed brain may be explained by different types of receptors represented in these brain structures. Particularly in hippocampus two types of receptors are identified—high affinity low capacity mineralocorticoid receptors (MR) and low-affinity high capacity glucocorticoid receptors (GR). First ones are working under basal non-stress or mild stress conditions, while second—under high-stress conditions. The hippocampus contains lots of high affinity receptors and is activated by moderate rises of cortisol, while the frontal lobe has only low affinity receptors and is activated later, or when cortisol reaches its peaks. This variability may be involved in salutary and

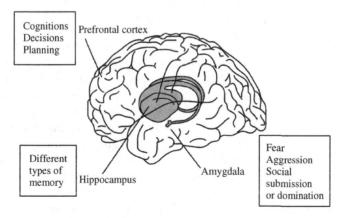

FIGURE 2.4 Main brain structures involved in stress reaction.

deleterious effects of stress and may underlie individual susceptibility to stress, depending on genetic-based receptors affinity and other features (Sapolsky, 2015). Hippocampus, prefrontal cortex and amygdala are also functionally different with regards of stress response: while hippocampus and the frontal cortex take part in shutting down HPA activity, amygdala, which is involved in fear processing, activates HPA. It coincides with differential neuroplasticity under stress in these brain structures. Thus, MR versus GR reactivity, as well as gene variants of these receptors seems to be crucial for regulation of coping, cognition, resilience, and vulnerability to stress (De Kloet et al., 2016).

Neuronal and molecular disturbances in the prefrontal cortex and other parts of the brain are triggered not only by cortisol, they may be also induced by CRH. CRH plays not only the role of regulator of the HPA, but also performs independent functions in the nerve tissue as a neurotransmitter (Snyder, Soumier, Brewer, Pickel, & Cameron, 2011). Kindling mechanisms also may be involved the formation of stress-induced mental health problems, when continuous stress effects "sway" systems and create prerequisites for wear and tear in the brain (Rozanov, Yemyasheva, & Biron, 2011). Neurochemical systems that are involved in neurotoxic effects of cortisol are diverse, as research is progressing their list is growing and includes except excitatory amino acids, oxytocin, CRH, different neurotrophins (especially BDNF), immunophilins (shaperones), and different other proteins (McEwen, 2012). In spite of this diversity, stress effects are mostly dependent on some nonspecific important molecules. For instance a glucocorticoid receptor cofactor neurophilin FKB5, while being encoded by specific gene variant, is associated with depression, anxiety, and PTSD (Provencal & Binder, 2015).

Pronounced functional and structural effects of stress on such critical brain structures as hippocampus, prefrontal cortex and, especially amygdala is considered to be the main pathological mechanism of an array of neuropsychiatric disorders, including depression, anxiety, addictions, and even schizophrenia. Exposure to psychosocial stress (as accumulation of negative life events) is known to be one of the strongest risk factors of different psychiatric disorders. Besides genes-to-environment interactions that are important for development of mental health problems, recently epigenetic effects are also widely discussed as important components of stress-induced psychopathologies (Provencal & Binder, 2015). The essence of epigenetics will be discussed a little further, but it is important to mention here that such effects depend very much on the timing of negative effect. There is a growing body of evidence that if both negative and positive stress may have a long-lasting consequences in case it coincides with a specific sensitive period in the life history of the organism or an individual. In this context effect of stress on behavior, cognition and probability of mental health disturbances throughout the life span is pivotal.

STRESS ACROSS THE LIFE-SPAN
AND SOCIAL FACTORS

Recent studies have accumulated a lot of evidence that stress in different periods of life, from intrauterine development period to older age, has differing consequences for mental health. This knowledge is collected both from animal studies and observations on humans and is based on the dynamics of brain maturation. Here we concentrate on human studies so far as mental health problems and disorders are known risk factors of suicide, which is the main subject of our discussion. Supporters of the concept of prenatal psychology are discussing to what extend intrauterine period of brain development is susceptible to influences derived from mother's feelings, thoughts, and emotions. It is interesting therefore that children whose mothers experienced psychological stress or had severe negative life events during pregnancy may develop different neurodevelopment problems. Maternal stress, anxiety, depression, and glucocorticoids treatment during pregnancy have been linked with lower birth weight or smaller for gestational age size of the baby (Hedegaard, Henriksen, Sabroe, & Secher, 1993; Kapoor, Petropoulos, & Matthews, 2008; Monk, Spicer, & Champagne, 2012). As the author of a widely cited review on this topic S. Lupien emphasizes, maternal stress, depression and anxiety have been associated with persistent increased basal HPA axis activity in the offspring at different ages, from 6 months to 10 years (Lupien, 2009). It leads further to unsociable and inconsiderate behaviors, attention deficit hyperactivity disorder and sleep disturbances, as well as depressive symptoms, drug abuse, and mood and anxiety disorders (Lupien, 2009).

Newborn babies are fully dependent on their mothers. During early period of breast feeding and further mother care actually plays the role of "external incubation," which is further substituted by more complex parental care. All together these interactions are producing an effect known as parental bonding. Here intrauterine risks already start to manifest themselves, especially if mother–baby interactions are developing in neglectful or affectionless way. The last also may have long-lasting consequences. In accordance with known facts about effects of stress hormones, especially cortisol, low birth weight in combination with low maternal care is associated with reduced hippocampal volume in adulthood (Buss et al., 2007). Further in development it is associated with an increased risk of behavioral problems and functional deregulation of HPA in response to everyday stressors (Albers, Riksen-Walraven, Sweep, & de Weerth, 2008). Most pronounced mental health problems and HPA disturbances are associated with child physical and sexual abuse as well as emotional torture or low maternal warmth. The last is especially typical for institutionalized children (Tarquis, 2006).

Next life period, which is commonly associated with high level of stress, is adolescence, a period of emotional and cognitive activation and emerging psychological problems. Here besides different types of abuse the role of poor economic factor is enhanced—adolescents who grew up in such poor conditions usually have higher baseline glucocorticoid levels, as do adolescents whose mothers were depressed in the early postnatal period (Lupien, 2009). Lower socio-economic conditions, except many health and high stress risks, may mean feeling of inferiority, frustration, low self-esteem, and social exclusion. Reduced hippocampal volume is not very common in case of adolescent's stress, by contrast, alterations in gray matter volume and the neuronal integrity of the frontal cortex and reduced size of the anterior cingulate cortex have been (Cohen et al., 2006). Adolescence is still a period of structural maturation of the brain, during which limbic regions myelination is outscoring cortical structures, and stress can seriously interfere with these processes (Romeo, 2016). Interestingly, one study reported that low self-esteem, a potent predictor of increased reactivity to stress in humans, is also associated with reduced hippocampal volume (Pruessner et al., 2005). It links together subjective feelings associated with social interactions and self-representation in the society and biological sequelae of stress.

Thus, stress across different periods of development and maturation can have differing consequences. However stressful experiences have a tendency of clustering, especially if an individual belongs to socially unprotected or stigmatized group. There is a high probability that early life stress may be associated with stress in childhood, than in adolescence, and further in life until old age. Each stressful experience may have consequences regarding susceptibility to further negative influences. Mental health at older age also appeared to be dependent to previous stages of development (Marin et al., 2011). Studies of adults, who suffered childhood abuse or extreme stressful situations when they were young, reveal hyperreactivity of the HPA axis in abused, depressed individuals, and hypoactivity in those with PTSD (Heim et al., 2000; Yehuda, Golier, & Kaufman, 2005).

Summing up, Lupien (2009) suggests the following model of stress effects on mental health across the lifespan. Stress in the prenatal period affects the development of many brain regions that are involved in regulation of the HPA axis—that is, the hippocampus, the frontal cortex, and the amygdala, executing programming effects, which will become transparent years later. Prenatal effects are the strongest and may lead to different mental health problems across the life span, compromising many critical brain structures. During childhood the hippocampus, which continues to develop after birth, seems to be the most vulnerable structure. By contrast, in adolescence the frontal cortex, which undergoes major development at this stage, may be most vulnerable to the effects of stress, possibly leading to a protracted glucocorticoid response to stress that persists into

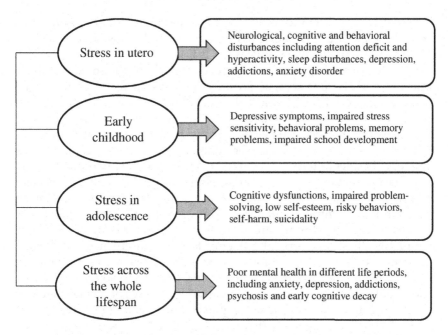

FIGURE 2.5 Timing of harmful stress and prevailing mental health consequences.

adulthood. In adulthood and old age the brain regions that undergo the most rapid decline as a result of aging appear to be highly vulnerable to the effects of stress hormones (Lupien, 2009). These data are summarized in Fig. 2.5.

Results of this comprehensive analysis suggest prevailing mental health problems in individuals with regards of the timing of stress. While prenatal stress predisposes to many disorders, from depression to schizophrenia, early postnatal stress may be risk factor for hyperactivity, and attention deficit disorder or, on the contrary, risk for development of the "inhibited child" phenotype. The latter further in life may develop into depression, anxiety, or addictions. Stress in adolescence is a risk factor for personality disorders, impaired decision making, cognitive abnormalities and, presumably, suicidality. Further life stress, from adulthood to older age, elevates chances of early cognitive decline and dementias. In reality all types of stress may be rather interconnected, especially if it comes about more global risk factors that belong to the social environment, like low socio-economic status, or living in a stressful environment with low level of public safety, or stressful family conditions due to dysfunctional family environment. Our study has found high association between early life stress (before 18) and adult life stress, which may be attributed also as a "personal high stress style" (Rozanov et al., 2011). Though lifetime stress may act according to different scenario (high, medium, ascending, or descending), trajectories of life stress appear more important than any single time-point in predicting associations with depression and anxiety (Herbison, Allen, Robinson, & Pennell, 2015).

Such thinking leads to more developed interdisciplinary models, which links together social factors, personality, life stress, perceived stress, neuroplasticity, and HPA regulation, thus combining a whole set of possible risk factors, mechanisms of vulnerability, and resilience and mental health outcomes. For instance, McEwen (2016) is drawing attention to the fact, that brain, as the central organ for adaptation to experiences, including stressors, besides altering systemic function through neuroendocrine, autonomic, immune, and metabolic stress-induced alterations, is also capable of changing its own architecture across the life span. So far as the brain is the central regulatory organ of behavior, alterations in brain function caused by chronic stress can have direct and indirect effects on cumulative allostatic overload. Systemic feedback mechanisms in chronic stress are the source of alterations in brain regions that are responsible for memory, cognitive functions, emotions, and self-regulation. All this may lead to unhealthy behavioral, emotional, and cognitive outcomes that add to negative stress effects. Therefore, plasticity-facilitating positive behavioral interventions that may attenuate negative effects are of great importance, though they still have to be better studied (McEwen, 2016). Another author who is discussing such integrative models, concentrate on the role of social environment (Holmes, 2016). She also starts from the point, that social environment sculpts the mammalian brain throughout life. Adult neurogenesis, the birth of new neurons in the mature brain, can be up- or down-regulated by various social manipulations. These include social isolation, social conflict, social status, socio–sexual interactions, and parent/offspring interactions. M. Holmes points on such factors as sex- and region-specific variations and argues that using a comparative approach to understand how adult-generated neurons influence social behaviors will shed light on how adult neurogenesis contributes to survival and reproduction (Holmes, 2016). Such modeling provides a background for integrative approach that puts together neuroendocrinology of stress, brain structural and functional dynamics, social context, behavior, emotions, cognition, and mental health. Timing of stressful events, programming of vulnerability or resilience, interactions between life stress, and biological underlying mechanisms suggest great importance of epigenetic mechanisms that are known to establish long-lasting patterns of genes activity.

Within the context of stress, its neurobiological correlates, neuroplasticity, and mental health consequences, one of the most intriguing aspects is perceived stress, the feeling of stress, stress "in the head" or mental stress, induced by fears, anxiety about future, negative thoughts or feelings. All these types of feelings are typical in modern life and are rather prevalent in young people and older adults, according to previously mentioned epidemiological studies. It remains not clear what type of stress is more harmful—real or perceived. In field studies about 60% of respondents show a moderate or severe perceived stress condition, with prevalence of females and young subjects. This perceived stress is associated with such stress-related parameters as prolactin level, consumption of coffee, chocolate, alcohol, and cigarettes smoking and is influenced by external urban physical/chemical and psychosocial factors (Tomei et al., 2012). It must be also taken into consideration that knowledge about stress and its deleterious effects have become very common. Stress as a concept (though not understood deeply by many people) became part of everyday communications. Worries about the fact that stress affects health is associated with worse health and mental health outcomes, those who reported that stress affected their health had an increased risk of premature death (Keller et al., 2012). Negative stress beliefs, that is, being convinced that "stress is bad for you," predict somatic symptoms in young adults (Fischer, Nater, & Laferton, 2016).

The subjective (psychological, emotional) stress experiences are assessed by self-reports of respondents, and there is a usual labeling of these experiences as less reliable or to a lesser extent credible. Nevertheless, there are many studies that confirm significant correlation between cortisol responses and perceived emotional stress (Campbell & Ehlert, 2012). There is an obvious link between physiological mobilization, emotions and cognitive evaluations which can be assessed only as subjective opinions. Returning to Lazarus definition and general understanding of stress in humans, role of a cognitive appraisal, and reappraisal appears to be crucially important. In a majority of situations stress is not just what happens, stress is what is understood as stress. For instance, in quasi experimental study of people who appeared in the situation of the terrorist attack in Oslo in 2011, perceived life threat in those who were only indirectly exposed to bomb explosion was associated with PTSD risk in the same manner as in those who exposed directly. Thus, just thinking that one's life is in danger may cause PTSD (Heir, Blix, & Knatten, 2016).

Within this context adaptive neural activations are associated with feeling of "being stressed" of fears about future, which is inherent to our perceptions and may lead to more active central neural processes. Accordingly, emerging correlates of maladaptive coping (hopelessness, anger, anxiety) are coupled with cognitive processes and with outcome expectances (Campbell & Ehlert, 2012). Sensory inputs integrated by cortical mechanisms and processed by brain circuits and provide cognitive evaluation of the significance and potential danger of the stimulus, as well as of coping resources. This leads to stimulation of emotional reactions in the limbic system, which in turn, promotes activation of neuroendocrinological mechanisms (Herman, Ostrander, Mueller, & Figueiredo, 2005). In case of chronic stress or as a result of early life programming all above mentioned brain structures may be compromised or hyperactive. This may lead to self-enhancement of perceived emotional stress. There are studies trying to objectify perceived stress, for instance, by measuring this subjective feeling and linking it to known stress-related pathologies, for instance, cardiovascular diseases. It is interesting, that correlations do exist, and quite expectedly they are culturally specific, for instance they are stronger in Russians, weaker in Americans and almost not evident in Chinese (Glei et al., 2013). It is also interesting to mention that talking about stress is rather common nowadays; the word stress becomes a part of a common cultural discourse, which is also dependent on the cultural context. Talking about stress, claiming stress or complaining on stressful life both helps to explain ones' own state and, possibly to cope due to social support. On the other hand but in some cases it may lead to enhancement of internal feeling of stress if social support is scarce (Pietilä & Rytkönen, 2008).

All this gives us a possibility to see how complex is the problem of stress and how many "layers" it has. The concept of stress have evolutionized from Selye's general adaptation syndrome, focused mostly on physiological processes, to psychic and psycho-emotional perceived stress concept, that encounters cognitive processes and coping skills. Modern studies on neuroendocrinology of stress, discovery of neuroplasticity, elucidation of the role of psychosocial factors in stress provide an interdisciplinary holistic view on the role of stress in health and disease, especially in mental health. Stress, especially in a form of psychosocial stress, is acknowledged as one of the most important risk factors of self-destructive behaviors, including self-mutilation, addictions, parasuicide and completed suicide (Wasserman & Sokolowski, 2016). Stress has gained a role of a global factor, while all above mentioned mental health problems are growing in response to modernization, global westernization and identity loss. Nevertheless in every population there are both those who may be very susceptible to stressors and those who remain resilient, as well as those who may even benefit from growing stress exposure. For instance, high neuroticism, impulsivity and aggressiveness may be adaptive in hostile and stressful environment and maladaptive in calm situations, clip thinking may be beneficial in the sphere of show-business and may lead to emergence of a new genius, but may lead to arrogant ignorance and anti-intellectualism in other individuals and to deleterious results in other spheres. The pressure of the environment reveals inequities, which may be supported by environments and lead to deeper differentiation.

The origin of vulnerability and resilience is traditionally linked to biological (genetic and epigenetic), psychological, and social factors. Recent studies have drawn attention to such already known for decades but undeservedly forgotten phenomena, as epigenetic transformations. Their role in the long-term programming of biological processes due to changed pattern of genes activity seem to be very relevant for understanding vulnerability and resilience to stress, and therefore variability of susceptibility to diseases and mental health problems. Their potential heritability or transgenerational transmission, which is not linked to changes in DNA sequences, both through behaviors and via germ-line cells suggest very interesting perspectives of study and explanations of the changing situation with suicides among younger generations. What is the nature of epigenetic transformations, how are they influenced by contextual factors and particularly by stress, can they be really inherited and does it mean recognition of heritability of the acquired experiences, what interventions can be suggested that may purposefully influence or even reverse these transformation—these and other questions will be subject of discussion in further chapters.

References

Adolphs, R., Tranel, D., & Damasio, A. R. (1998). The human amygdala in social judgment. *Nature, 393*, 470–474.

Aguilera, G., Kiss, A., Liu, Y., & Kamitakahara, A. (2007). Negative regulation of corticotropin releasing factor expression and limitation of stress response. *Stress, 10*, 153–161.

Ahola, K., Sirén, I., Kivimäki, M., Ripatti, S., Aromaa, A., Lönnqvist, J., & Hovatta, I. (2012). Work-related exhaustion and telomere length: a population-based study. *PLOS One, 7*(7), e40186.

Albers, E. M., Riksen-Walraven, J. M., Sweep, F. C., & de Weerth, C. (2008). Maternal behavior predicts infant cortisol recovery from a mild everyday stressor. *Journal of Child Psychology and Psychiatry, 49*, 97–103.

Anderson, R. (2004). Stress and mental health. *Journal of the Royal Society for the Promotion of Health, 124*, 112–113.

Antoni, M. H. (2003). Stress-management effects on psychological, endocrinological and immune functioning in men with HIV infection: empirical support for psychoneuroimmunological model. *Stress, 3*, 173–188.

Belkic, K., Landsbergis, P. A., Schnall, P. L., & Baker, D. (2004). Is job strain a major source of cardiovascular disease risk? *Scandinavian Journal of Work Environment and Health, 30*, 85–128.

Binfa, L., Castelo-Branco, C., Blumel, J. E., Cancelo, M. J., Bonilla, H., Munoz, I., Vergara, V., Izaquirre, H., Sarra, S., & Rios, R. V. (2004). Influence of psycho-social factors on climacteric symptoms. *Maturitas, 48*, 425–431.

Buss, C., Lord, C., Wadiwalla, M., Hellhammer, D. H., Lupien, S. J., Meaney, M. J., & Pruessner, J. C. (2007). Maternal care modulates the relationship between prenatal risk and hippocampal volume in women but not in men. *Journal of Neurosciences, 27*, 2592–2595.

Campbell, J., & Ehlert, U. (2012). Acute psychosocial stress: does the emotional stress response corresponds with physiological responses. *Psychoneuroendocrinology, 37*, 1111–1134.

Cerel, J., Maple, M., van de Vende, J., Moore, M., Falherty, C., & Brown, M. (2016). Exposure to suicide in the community: prevalence and correlates in one U.S. state. *Public Health Reports, 131*, 100–107.

Charmandari, E., Tsigos, C., & Chrousos, G. P. (2005). Endocrinology of the stress response. *Annual Reviews in Physiology, 67*, 259–284.

Chida, Y., Hamer, M., Wardle, J., & Steptoe, A. (2008). Do stress-related psychosocial factors contribute to cancer incidence and survival? *Nature Clinical Practice Oncology, 5*, 466–475.

Cohen, S. (2005). The Pittsburg common cold studies: psychosocial predictors of susceptibility to respiratory infection illness. *International Journal of Behavioral Medicine, 12*, 123–131.

Cohen, S., Kamarck, T., & Mermelstein, R. (1983). A global measure of perceived stress. *Journal of Health and Social Behavior, 24*, 385–396.

Cohen, R. A., Grieve, S., Hoth, K. F., Paul, R. H., Sweet, L., Tate, D., Gunstad, J., Stroud, L., McCaffery, J., Hitsman, B., Niaura, R., Clark, C. R., McFarlane, A., Bryant, R., Gordon, E., & Williams, L. M. (2006). Early life stress and morphometry of the adult anterior cingulate cortex and caudate nuclei. *Biological Psychiatry, 59*, 975–982.

Danielsson, M., Heimerson, I., Lundberg, U., Perski, A., Stefansson, C. -G., & Akerstebdt, T. (2012). Psychosocial stress and health problems. Health in Sweden: The National Health Report 2012. Chapter 6. *Scandinavian Journal of Public Health, 40*(Suppl. 9), 121–134.

De Kloet, E. R., Otte, C., Kumsta, R., Kok, L., Hillegers, M. H., Hasselman, M. H., Kliegel, D., & Joels, M. (2016). Stress and depression: a crucial role of mineralocorticoid receptor. *Journal of Neuroendocrinology, 28*, https://www.ncbi.nlm.nih.gov/pubmed/26970338.

Dedovic, K., Duchesne, A., Andrews, J., Engert, V., & Pruessner, J. C. (2009). The brain and the stress axis: the neural correlates of cortisol regulation in response to stress. *NeuroImage, 47*, 864–871.

Fischer, S., Nater, U. M., & Laferton, J. A. (2016). Negative stress beliefs predict somatic symptoms in students under academic stress. *International Journal of Behavioral Medicine, 18*, 1–6.

Glei, D. A., Goldman, N., Shkolnikov, V. M., Jdanov, D., Shkolnikova, M., Vaupel, J. W., & Weinstein, M. (2013). Percieved stress and biological risk: is the link stronger in Russians than in Taiwanese and Americans? *Stress, 16*, 411–420.

Gollenberg, A. L., Liu, F., Brazil, C., Drobnis, E. Z., Guzick, D., Overstreet, J. W., Redmon, J. B., Sparks, A. E. T., Wang, C., & Swan, S. H. (2010). Semen quality in fertile men in relation to psychosocial stress. *Fertility and Sterility, 93*, 1104–1111.

Goyal, S., Jajoo, S., Nagappa, G., & Rao, G. (2011). Estimation of relationship between psychosocial stress and periodontal status using serum cortisol level: a clinico-biochemical study. *Indian Journal of Dental Research, 22*, 6–9.

Hall, E., & Burt, V. K. (2012). Male fertility: psychiatric considerations. *Fertility and Sterility, 97*, 434–439.

Hayashi, R., Wada, H., Ito, K., & Adcock, I. M. (2004). Effects of glucocorticoids on gene transcription. *European Journal of Pharmacology, 500*, 51–62.

Hedegaard, M., Henriksen, T. B., Sabroe, S., & Secher, N. J. (1993). Psychological distress in pregnancy and preterm delivery. *British Medical Journal, 307*, 234–239.

Heim, C., Newport, J., Heit, S., Graham, Y. P., Wilcox, M., Bonsall, R., Miller, A. H., & Nemeroff, C. B. (2000). Pituitary-adrenal and autonomic responses to stress in women after sexual and physical abuse in childhood. *Journal of American Medical Association, 284*, 592–597.

Heir, T., Blix, I., & Knatten, C. K. (2016). Thinking that one's life was in danger: perceived life threat in individuals directly or indirectly exposed to terror. *British Journal of Psychiatry, 209*(4), 306–310.

Herbison, C., Allen, K., Robinson, M., & Pennell, C. (2015). Trajectories of stress events from early life to adolescence predict depression, anxiety and stress in young adults. *Psychoneuroendocrinology, 61*, 16–17.

Herman, J. P., Ostrander, M. M., Mueller, N. K., & Figueiredo, H. (2005). Limbic system mechanisms of stress regulation: hypothalamo-pituitary-adrenocortical axis. *Progress in Neuro-Psychopharmacology and Biological Psychiatry, 29*, 1201–1213.

Herman, J. P., McKlveen, J. M., Ghosal, S., Kopp, B., Wulsin, A., Makintosh, R., Scheiman, J., & Myers, B. (2016). Regulation of the hypothalamic-pituitary-adrenocortical stress response. *Comprehensive Physiology, 6*, 603–621.

Holmes, M. M. (2016). Social regulation of adult neurogenesis: a comparative approach. *Frontiers of Neuroendocrinology, 41*, 59–70.

Inglehart, R. (1997). *Modernization and postmodernization: cultural, economic, and political change in 43 societies.* Princeton, NJ: Princeton University Press.

Johansen, C. (2012). Stress and cancer. *Ugeskrifr for Laeger, 174,* 208–210.

Kamarck, T. W., Muldoon, M., Shiffman, S., Sutton-Tyrell, K., & Gwaltney, C. J. (2004). Experiences of demand and control in daily life as correlates of subclinical carotid atherosclerosis in a healthy older sample: the Pittsburg Healthy Heart Project. *Health Psychology, 23,* 24–32.

Kapitsa, S. P. (2004). Ob uskorenyi istoricheskogo vremeni (On enhancement of the historical time). *Novaya i Noveyshya Istoriya (New and Newest History), 6,* Available from: http://vivovoco.astronet.ru/VV/JOURNAL/NEWHIST/KAPTIME.HTM.

Kapoor, A., Petropoulos, S., & Matthews, S. G. (2008). Fetal programming of hypothalamic-pituitary-adrenal (HPA) axis function and behavior by synthetic glucocorticoids. *Brain Research Reviews, 57,* 586–595.

Katerndahl, D. A., & Parchman, M. (2002). The ability of the stress process model to explain mental health outcomes. *Comprehensive Psychiatry, 43,* 351–360.

Keller, A., Litzelman, K., Wisk, L. E., Maddox, T., Cheng, E. R., Creswell, P. D., & Witt, W. P. (2012). Does the perception of stress affects health matter? The association with health and mortality. *Health Psychology, 31,* 677–684.

King, A. (2016). Rise of resilience. *Nature, 531,* S18–S19.

Kirchengast, S., & Haslinger, B. (2010). The association between mild geriatric depression and reproductive history—a Darwinian approach. *Anthropologischer Anzeiger, 68,* 209–220.

Kumar, R., & Thompson, E. B. (2005). Gene regulation by the glucocorticoid receptor: structure/function relationship. *Journal of Steroids Biochemistry and Molecular Biology, 94,* 383–394.

Lazarus, R., & Folkman, S. (1984). *Stress, appraisal, and coping.* NY: Springer Publishing Company.

Lundberg, U. (2006). Stress, subjective and objective health. *International Journal of Social Welfare, 15,* S41–S48.

Lupien, S. (2009). Effects of stress throughout lifespan on the brain, behavior and cognition. *Nature Reviews Neuroscience, 10,* 434–445.

Marin, M. F., Lord, C., Andrews, J., Juster, R. P., Sindi, S., Aresenault-Lapierre, G., Fiocco, A. J., & Lupien, S. J. (2011). Chronic stress, cognitive functioning and mental health. *Neurobiology of Learning and Memory, 96,* 583–595.

Martikainen, P., Bartley, M., & Lahelma, E. (2002). Psychosocial determinants of health in social epidemiology. *International Journal of Epidemiology, 31,* 1091–1093.

McEwen, B. (1993). Stress and the individual. *Archives of Internal Medicine, 153,* 2093–2101.

McEwen, B. C. (1999). Stress and the aging hippocampus. *Frontiers in Neuroendocrinology, 20,* 49–70.

McEwen, B. (2012). Brain on stress: how the social environment gets under the skin. *Proceedings of National Academy of Sciences, 109*(Suppl. 2), 17180–17185.

McEwen, B. S. (2016). In pursuit of resilience: stress, epigenetics, and brain plasticity. *Annals of the New York Academy of Sciences, 1373*(1), 56–64.

Meaney, M. J., Szyf, M., & Seckl, J. R. (2007). Epigenetic mechanisms of perinatal programming of hypothalamic pituitary-adrenal function and health. *Trends in Molecular Medicine, 13,* 269–277.

Michmizos, D., Koutsouraki, E., Asprodini, E., & Baloyannis, S. (2011). Synaptic plasticity: a unifying model to address some persisting questions. *International Journal of Neuroscience, 121,* 289–304.

Monk, C., Spicer, J., & Champagne, F. A. (2012). Linking prenatal maternal adversity to developmental outcomes in infants: the role of epigenetic pathways. *Development and Psychopathology, 24,* 1361–1376.

Park, B. C., Im, J. S., & Ratcliff, K. S. (2014). Rising youth suicide and changing cultural context in South Korea. *Crisis, 35,* 102–109.

Pietilä, I., & Rytkönen, M. (2008). Coping with stress and by stress: Russian men and women talking about transition, stress and health. *Social Science and Medicine, 66*, 327–338.

Podolska, M. Z., & Bidzan, M. (2011). Infertility as a psychological problem. *Ginekologia Polska, 82*, 44–49.

Provencal, N., & Binder, E. B. (2015). The neurobiological effect of stress as contributors to psychiatric disorders: focus on epigenetics. *Current Opinion in Neurobiology, 30*, 31–37.

Pruessner, J. C., Baldwin, M. W., Dedovic, K., Renwick, R., Mahani, N. K., Lord, C., Meaney, M., & Lupien, S. (2005). Self-esteem, locus of control, hippocampal volume, and cortisol regulation in young and old adulthood. *Neuroimage, 28*, 815–826.

Revesz, D., Milaneschi, V., Terpstra, E. M., & Penninx, B. W. (2016). Baseline biopsychosocial determinants of telomere length and 6-year attrition rate. *Psychoneuroendocrinology, 67*, 153–162.

Romeo, R. D. (2016). The impact of stress on the structure of adolescent brain: implications for adolescent mental health. *Brain Research*, https://www.ncbi.nlm.nih.gov/pubmed/27021951.

Rozanov, V. A., Yemyasheva, Zh. V., & Biron, B. V. (2011). Influence of childhood traumatic experience on general life stress accumulation and suicidal tendencies through the lifespan. *Ukrainian Medical Journal, 6*, 94–98.

Sapolsky, R. (2004). *Why zebras don't get ulcer. The acclaimed guide to stress, stress-related disease and coping* (3rd ed.). NY: Henry Holt and Co.

Sapolsky, R. (2015). Stress and the brain: individual variability and inverted U-curve. *Nature Neuroscience, 18*, 1344–1346.

Sapolsky, R., Krey, L., & McEwen, B. S. (1986). Neuroendocrinology of stress and aging: the glucocorticoid cascade hypothesis. *Endocrine Reviews, 7*, 284–301.

Schwartz, S., & Meyer, I. H. (2010). Mental health disparities research: the impact of within and between groups analyses on tests of social stress hypothesis. *Social Science and Medicine, 70*, 1111–1118.

Schwartz, J., Evers, A. W., Bundy, C., & Kimball, A. B. (2016). Getting under the skin: report from the international psoriasis council workshop. *Frontiers in Psychology, 7*, 87.

Semenovskikh, T. (2014). The phenomenon of «clip-thinking» in the educational high school environment. *Naukovedenie, 5*(24), http://naukovedenie.ru.

Siegrist, J. (1996). Adverse health effects of high effort/low reward conditions. *Journal of Occupational Health Psychology, 1*, 27–41.

Snyder, J. S., Soumier, A., Brewer, M., Pickel, J., & Cameron, H. A. (2011). Adult hippocampal neurogenesis buffers stress responses and depressive behavior. *Nature, 476*, 458–461.

Surdea-Blaga, T., B ban, A., & Dumitrascu, D. L. (2012). Psychosocial determinants of irritable bowel syndrome. *World Journal of Gastroenterology, 18*, 616–626.

Tamashiro, K. L., Sakai, R. R., Shively, C. A., Karatsoreos, I. N., & Reagan, L. P. (2011). Chronic stress, metabolism, and metabolic syndrome. *Stress, 14*, 468–474.

Tarquis, N. (2006). Neurobiological hypothesis relating to connections between psychopathy and childhood maltreatment. *Encephale, 32*, 377–384.

Theorell, T., & Karasek, R. A. (1996). Current issues relating to psychosocial job strain and cardiovascular disease research. *Journal of Occupational Health and Psychology, 1*, 9–26.

Tomei, G., Sancini, A., Capozzella, A., Caciari, T., Tomei, F., Nieto, H. A., Gioffrè, P. A., Marrocco, M., De Sio, S., Rosati, M. V., & Ciarrocca, M. (2012). Perceived stress and stress-related parameters. *Annali di igiene: medicina preventiva e di comunità, 24*, 517–526.

Tosevski, D. L., & Milovancevic, M. P. (2006). Stressful life events and physical health. *Current Opinion in Psychiatry, 19*, 184–189.

Tsigos, C., & Chrousos, G. P. (2002). Hypothalamic-pituitary-adrenal-axis, neuroendocrine factors and stress. *Journal of Psychosomatic Research, 53*, 865–871.

Vyas, A., Mitra, R., Shankaranayana, R. B. S., & Chattarji, S. (2002). Chronic stress induces contrasting patterns of dendritic remodeling in hippocampus and amygdaloid neurons. *Journal of Neurosciences, 22*, 6810–6818.

Wasserman, D., & Sokolowski, M. (2016). Stress-vulnerability model of suicidal behaviours. In D. Wasserman (Ed.), *Suicide. An unnecessary death* (2nd ed., pp. 27–37). NY: Oxford University Press.

Watanabe, N., Stewart, R., Jenkins, R., Bhugra, D. K., & Furukawa, T. A. (2008). The epidemiology of chronic fatigue, physical illness, and symptoms of common mental disorder: a cross-sectional survey from the second British National Survey of Psychiatric Morbidity. *Journal of Psychosomatic Research, 64,* 357–362.

Wilkinson, R., & Pickett, K. (2010). *The spirit level. Why greater equality makes societies stronger.* NY: Bloomsbury Press.

Wiltink, J., Beutel, M. E., Till, Y., Ojeda, F. M., Wild, P. S., Munzel, T., Blankenberg, S., & Michal, M. (2011). Prevalence of distress, comorbid conditions and well-being in the general population. *Journal of Affective Disorders, 130,* 429–437.

Woon, F. L., Sood, S., & Hedges, D. W. (2010). Hippocampal volume deficits associated with exposure to psychological trauma and posttraumatic stress disorder in adults: a meta-analysis. *Progress in Neuro-Psychopharmacology & Biological Psychiatry, 34,* 1181–1188.

World Health Organization (2008). *The Global Burden of Disease,* Update.

Yehuda, R., Golier, J. A., & Kaufman, S. (2005). Circadian rhythm of salivary cortisol in Holocaust survivors with and without PTSD. *American Journal of Psychiatry, 162,* 998–1000.

3

What Is Epigenetics?
Is It Transgenerational?

In the first decades of 20th century, famous Russian scientist and Nobel Prize winner for his studies on physiology of digestion Ivan Petrovich Pavlov started a new wave of research on conditioned reflexes and higher nervous (mental) activity. His experiments provided very convincing data about formation and fading of reflexes, and though intrinsic cellular mechanisms that "linked together" different parts of sensory systems and cortical integrative functions remained at that time unclear, he was discussing possible ways how long-lasting effects may develop. At the physiological congress in Groningen in 1913 he expressed the idea that in the long run well-developed and periodically supported conditioned reflexes can be transmitted in generations, that is, become congenital. At that moment he was basing on the results of his collaborator Nikolay Studentsov who started studies with conditioned reflexes in white mice teaching them to run to the feeder box after the ringtone. In this case Pavlov changed his traditional and well-studied experimental animal—the dog, and turned to an animal with the short generative cycle. It is not known exactly, but there is a very high probability that for experiments adult mature male animals were used—it was a tradition at that time. In a period from 1921 to 1923, five generations of mice were taught to come up to the feeder after the sound of electric bell. Initial number of combinations of conditioned (electrical bell) and unconditioned stimuli (food) to produce stable effect was 298, in the second generation it was already 114, and in the fifth generation it was enough to make only 6 combinations of the stimuli. Pavlov was very interested in this result and mentioned it in his article in *Science* (Pawlow, 1923). Later he established a department of Experimental Genetics of the Higher Mental Activity in his new experimental institute in Koltushy near Leningrad (today Saint-Petersburg). He was already discussing importance that inheritable character of conditioned reflexes may have for child development and

Stress and Epigenetics in Suicide. http://dx.doi.org/10.1016/B978-0-12-805199-3.00003-8

school education. It is known that shortly after his publication in *Science* leading geneticists Nikolay Konstantinovich Koltsov and Thomas Hunt Morgan have strongly opposed Pavlov and criticized his work from the point of view of classical genetics. Further studies by different authors did not confirm transmission of patterns acquired during training in animals when classical conditioned reflexes methodology was used. Pavlov asked another collaborator Professor Ganike, who was very keen in technical standardization of experiments, to check results of Studentsov, and soon had to confess that results were not confirmed. Nevertheless, it happened so that results of Ganike were not published, while several other papers with actually confirmed inheritance of some acquired characteristics in mammals (guinea pigs) and crustaceans were, but remained almost unknown to physiologists and psychologists. Severe criticism from classical geneticists and unconvincing experimental results finally made Pavlov change his opinion and he even published a letter in "Pravda" in which he stated that he is not supporting the idea of transgenerational transmission of acquired conditioned reflexes any longer. The fact that official Soviet newspaper was chosen for this gives an impression how important this letter was for Pavlov. It is said that this episode prevented Ivan Pavlov from receiving his second Nobel Prize, for which he was nominated several times and for winning which he had rather high chances as a world renowned leader of physiologists and a pioneer in higher mental activity studies at the moment.

And now in connection with this story it is quite appropriate to mention results of rather recent study of the group from USA, in which researchers studied long-term potentiation in brains of female 15 days old (preadolescent) mice that had a genetically created defect in memory (knocked-out and lacking Ras-GRF genes, a family of guanine nucleotide exchange factors mediating NMDA receptors, which are involved in Ca-dependent processes of memory in hippocampus). When these young mice were exposed to enriched environments, including plenty of stimulatory objects in the cage, enhanced social interactions, and voluntary exercise for 2 weeks, the memory impairment was reversed, which gives an impression how environmental influences can help to overcome a genetic defect. This is not a surprise in view of recent knowledge about huge potential of neuroplasticity, but this is not the end of the story. When in few months, the same well brought-up mice were fertilized, their offspring (that had of course the same genetic mutation, transmitted in a Mendelian way) did not show any indication of the memory defect even though they were never exposed to an enriched environment like their mothers! Thus, the enriched environment, which means a variety of external signals, in a specific time window (when animals are rather young) can cause changes that are transmitted to the next generation (Junko, Shaomin, Hartley, & Feig, 2009). The authors are very clearly stating that this effect may be referred as "Lamarckian" phenomenon, the type of events which seemed to have been condemned and rejected by science irrevocably and latent adherence to which cost Pavlov his second Nobel Prize.

Almost at the same time as Pavlov was discussing possible heredity of learned experiences, another Russian, an agronomist Trofim Denisovitch Lysenko, was experimenting with plants, particularly with wheat, studying the influence of cold exposure on acceleration of flowering. He was achieving practical results that were important for Soviet Russia where after civil war economic collapse was accompanied with hunger, high death rates, and other adversities. Soon after his results showed that exposure of seeds to humidity and further to cold in a certain time window enhanced flowering and provided better harvesting in regions with severe climate, millions of tons of wheat seed were treated by Lysenko's method. Moreover, he developed his own genetic theory which sometimes is characterized as "naïve Lamarckism" (Graham, 1987). In this system the emphasis was made not on genes conservatism, but on the influence of ecological factors—nutrients, light, water, external gases, presence of weeds, relations between the plants, and so on, and changes in the plant (or animal) organism that were induced by these factors. The central idea of his concept was that heredity "is being developed during whole life of the organism." In his perception the environmental influences could completely change the nature of the plant and this quality may become a stable trait. Moreover, he was differentiating between several types of heredity among which Mendelian heredity was not the most important one. Observations of Lysenko led to him to understanding that the whole life of the plant can be divided into several distinct periods, each period demanding different nutritional and other external factors, while transitions between these periods were exceptionally susceptible to external influences (Zhivotovsky, 2014).

His attempts to change the plant organism with the use of external influences (to mentor the plant) were opposing classical genetics, which ignited vivid and sometimes dangerous discussions. Lysenko's positions were preferential because he suggested immediate practical results with economic effect, while classical geneticists were mostly theorizing. Besides, he was supported by official ideology of those times—revolutions will change the world and a human being, not only plants. Thus, Lysenko has made a great career and gained a lot of administrative power. Soon the conflict between ideas turned into the persecution of classical geneticists (that were announced "weissmanists" and "morganists" who were an obstacle on the way of achieving practical results) by proponents of Lysenko's new science—agrobiology. In 1948–52 in the USSR after the report of Lysenko on the session of the Academy of Agricultural sciences, several landmark sessions in all Academies (general, medical, and pedagogical) took place, during which Pavlov's ideas were rehabilitated on the basis of Lysenko's ideology, and many classical geneticists who were opposing him were expelled from university departments and laboratories. Later, in the end of 1950s when political situation changed, Lysenko had lost most of his administrative positions, the revenge of classical geneticists was also severe—the whole biological system and genetic views of Lysenko were announced a pseudoscience and were abandoned. And as it turns out, unfortunately, because some of his observations are much better understood today within the context of the role of epigenetics in development of the living organism and transgenerational transmission of epigenetic marks and phenotypes.

Today, mechanisms of vernalization (jarovization as Lysenko called it) are quite well understood; they are linked to DNA methylation/demethylation and epigenetic effects of silencing and expression of certain genes, which are providing "on–off" mechanisms of metabolic processes in the corn. Cooling of the seed in a specific time frame on the stage of germination makes it "remember" the ecological exposure; it is an important process that has practical applications (Amasino, 2004). Moreover, the idea that the trait, acquired by the organism during the life-span as a result of external exposures, may be transferred to subsequent generations does not at all look like heresy. It is quite recently that in a very serious journal an extensive review summarized about 100 well-established cases of transgenerational inheritance of environmentally induced phenotypes in different biological species—from bacteria to higher plants and mammals. This heredity can be well noticed in several generations—from three to few dozen but may fade with time, if environment does not support it, or due to other reasons (Jablonka & Raz, 2009).

So, what is going on? Is it reestablishment of Lamarck legacy? What is epigenetics, so frequently mentioned in modern biomedical and even psychiatric and psychological publications? Why it is so important for us with regard to the very human problem of suicide? We will try to outline briefly most modern trends and concepts in this field until we go further in our attempt to integrate knowledge on stress, genetics, epigenetics, and suicide.

EPIGENETICS—NEW LIFE OF OLD IDEAS AND MODERN DEFINITIONS

From its birth, the concept of epigenetics was of course very well known among specialists (for instance, developmental geneticist and evolutionary biologists), but was not very much popular in medical, psychological, behavioral, and psychiatric circles. Now we see the renaissance of this knowledge and its' influence in adjacent fields, which gives an impression of the revolution in science. But is it really a revolution? Epigenetic transformations started to be extensively discussed in the last decades because it became evident that they are involved in such basic phenomena as aging, cancer, health and diseases, stress vulnerability and resilience, experiences consolidation and memory formation, effects of ecological pollutants, small doses of ionizing irradiation, mechanisms of action of nutrients, and mental and behavioral disorders (Szyf, 2009b; Meaney & Szyf, 2010; Vaiserman, 2015a). Understanding that our appearance, behavior, stress responses, disease susceptibility, and longevity are linked to epigenetic transformations ignited huge interest to this type of events. Epigenetics suggests just a new way of thinking and a new scientific tradition in the field of physiology of humans and, of course, in medical sciences. At the same time, epigenetics is a reincarnation of old ideas about the nature of basic biological mechanisms associated with organism conceiving and development.

The term "epigenetic," that is, literally "over genetics, over heredity, above heredity," was used repeatedly in various contexts, including even in psychology (the epigenetic concept of development of the identity of Eriksson). This term goes back to the concept of "epigenesis" which means existence of a certain development plan, that is, when formation of complex forms is based on the initial simple basis. Ideas of epigenesis, which are believed to emerge in antiquity, resist to the concept of preformationism, according to which development means just growth (deployment) of the forms, which are already available in a germ. Preformation theory was prevailing until the 18th century, but with the development of microscopy it has faded so far as no material objects were revealed (like homunculi in the sperm). Nevertheless, the dispute between these theories did not become less intense; it simply reached the molecular and submolecular level and today looks more like a collision of philosophical concepts.

It is acknowledged that the term "epigenetics" was introduced to the biological terminology by British embryologist and geneticist Conrad Waddington in the middle of 20th century. It was the time when mechanisms of transmission of genes from generation to generation were already quite well understood, while action of genes during development from fertilized egg to a mature organism was understood rather poorly (Holliday, 1987). How pluripotent zygote gives rise to a variety of differentiated cell types remained obscure. It was clear that it was based on the same initial set of genes, but what made cells so different? Geneticists were thinking about different mechanisms, including restructuring of chromosomes and even loss of certain genes in different cell lines. Waddington's epigenetic hypothesis has firmly established the opinion that genes are remaining "on their usual places," but their activity in different stem cells clones becomes specified—some genes are silenced, while others remain active. This specific profile of genes expression makes future generations of differentiating cells unique. The longer way cell travels from the pluripotent zygote to terminally differentiated form belonging to specific tissue or organ, the more genes are silenced, though on this way some external clues may change the way by which this or other cell clone goes. To explain better how it worked, Waddington proposed a model of the "epigenetic landscape" and a ball that goes down the hill choosing different routes due to external cues as an imagined composition of multiple variants of development. The choice among these variants is determined by different signals that can be transmitted to the genetic material via biologically active components, mostly proteins, but also metabolites, hormones, and nutrients. It is important that this profile for each cell type is established once during early stages of development and remains the same for all future cell divisions (Pennisi, 2001). These mechanisms eventually lead to formation of about 270 terminally differentiated cell types which form human body and which remain morphologically and functionally stable during the whole life-span (until some tragic event induce transformation).

It is often mentioned that Waddington's concept can be defined as the "mechanism of interaction of genes and their products," which can also be understood as mechanism by which genes are responding to various intrinsic signals dependent on genetic program of development. It is much more interesting to what extent extrinsic or environmental signals (with respect to an embryo or a fetus, or a newborn or even mature organism) can influence development through biologically active components like proteins, hormones, or metabolites, which change their concentration due to external environment cues (Golubovskii, 2000). Indeed, it is known for centuries that intrauterine period of development is highly sensitive to toxic or stressful influences that may lead to different negative effects. On the contrary, we are not speaking about serious pathological development which touches structure and function of organs and systems of the organism and which can result in teratogenic effects that become evident at birth. Influences that can alter genes activities through epigenetic effects and that may result in subtle structural abnormalities (of brain, neuroendocrine system, or of other systems and organs) may become evident functionally after a very long period of time, maybe decades in case of humans, again as a result of interaction with the environment. Until recently it was not evident that some of the common age-dependent diseases have their roots in the intrauterine period or in the early period of postnatal life.

Moreover, there is a growing body of evidence that some environmentally induced phenotypes can be transmitted to subsequent generations, not only cell generations of the same body, but to progeny of the living organism (Richards, 2006). Such facts were known for many years to geneticists, but were interpreted in different terms and were referred as poorly understood events with heritable component (Holliday, 2006). Recently, these scattered facts are brought together and are receiving logical interpretation. One kind of events is those that are induced by extrinsic signals acting on the early stages of development, conserved through epigenetic mechanisms, and realized further during the life-span. Another type of events is effects of early harmful or contrary, stimulating exposures and experiences, that happen in the early stage of life and then appear in progeny, persist in several generation, and then fade. The third type of events is those that may happen to a grown-up organism and may change memory, aging, or cause cancer or mental disorder.

So, while main sphere of action of the epigenetics according to initial concept of Waddington was the period of embryogenesis, modern epigenetic science has extended the period of high probability of these events from embryogenesis to early life periods, and with regard to the cell biology of brain that accumulates experiences—to the whole life-span of the organism. Today, it becomes clear that on the basis of the same genome the human body develops not only hundreds of epigenomes, which determine stable cellular diversity, but also maybe myriads of epigenomes inside such organ as brain, where neurons are constantly reacting to stimuli and where structural and functional relations between neurons are adjusted by existing challenges and rewards, or in the immune system, which reacts to all external influences. As a result, the concept of epigenome has emerged that is understood in modern terms as a dynamic interface between nature and nurture (Tammen, Friso, & Choi, 2013; Szyf, 2009a). Definitions of epigenetics and epigenome are linked today to the concept of "soft inheritance"—a set of molecular mechanisms that allow environmental signals to establish reversible imprints (marcs) on the genetic material and to transmit these marcs further in several generations, both cell generations and progeny, but not in a rigid and strongly determined way, that is, these imprints can be reversed under certain conditions.

In view of these fast changes in biological thinking it is interesting to compare historical and modern definitions of epigenetics (Table 3.1). It is especially important so far as the word "epigenetics" is often used too loosely and it is necessary to remember that epigenetics and epigenetic inheritance are different concepts (Jablonka & Raz, 2009).

As can be seen from the table, there is an evolution of definitions from rather generalized "processes by which genotype gives rise to phenotype" to more focused "phenotypic variations that do not stem from variations in DNA base sequences and are transmitted to subsequent generations of cells or organisms." Half of definitions are negative, that is, they stress that epigenetics is based on molecular events that are not touching DNA base sequences. It is therefore necessary to give a brief description of these events on the biochemical level and to outline mechanisms of epigenetic inheritance, that is, how these molecular events can pass through mitosis and meiosis. Most sophisticated molecular mechanisms that surround these events are far beyond the scope of this book; actually, they can be

TABLE 3.1   Different Definitions of Epigenetics in a Historical Perspective

No.	Definition	Sources
1	Study of the processes by which genotype gives rise to phenotype	Waddinhton (1942) (cited by Wu & Morris, 2001)
2	The branch of biology which studies the causal interaction between genes and their products, which brings the genotype into being	Waddington (1942) (cited by Goldberg, Allis, & Bernstein, 2007)
3	Study of the changes in gene expression which occur in organisms with differentiated cells, and the mitotic inheritance of given patterns of gene expression	Holliday (1994)
4	Nuclear inheritance which is *not* based on differences in DNA sequence	Holliday (1994)
5	The study of changes in gene function that are mitotically and/or meiotically heritable and do *not* entail a change in DNA sequence	Wu and Morris (2001)
6	The study of stable alterations in gene expression potential that arise during development and cell proliferation	Jaenisch and Bird (2003)
7	Heritable changes in gene expression that *cannot* be tied to genetic variation	Richards (2006)
8	The study of any potentially stable and, ideally, heritable change in gene expression or cellular phenotype that occurs *without* changes in Watson–Crick base-pairing of DNA	Goldberg et al. (2007)
9	An epigenetic trait is a stably heritable phenotype resulting from changes in a chromosome *without* alterations in the DNA sequence	Berger, Kouzarides, Schickhattar, and Shilatifard (2009)
10	The study of the processes that underlie developmental plasticity and canalization and that bring about persistent developmental effects in both prokaryotes and eukaryotes	Jablonka and Raz (2009)
11	Epigenetic inheritance … occurs when phenotypic variations that *do not* stem from variations in DNA base sequences are transmitted to subsequent generations of cells or organisms	Jablonka and Raz (2009)

deeply understood only by those who are specializing in modern molecular genetics. On the contrary, any medical or psychological specialist, who has been educated in the classical tradition, may need an objective picture of these events (as the author has felt himself when was first confronted by a flow of studies and evidences of new mechanisms).

MOLECULAR MECHANISMS OF EPIGENETICS AND EPIGENETIC INHERITANCE—A SHORT REVIEW

All acquired and reversible molecular signatures that are involved in regulation of the genome expression are referred as epigenome, and whole complex of epigenetic marks dynamics and studies of epigenome are known recently as epigenomics. The central part of this knowledge is that DNA is not a stable molecule; on the contrary, it is being transformed constantly. Dynamics of the epigenome is determined by biochemical reactions controlled by quite definite enzymes. Several molecular mechanisms that target long-lasting programming of the gene expression that can be transmitted to next generations of cells and organisms have been identified. They are the following: (1) methylation of DNA molecules by cytosine residues with formation of the so-called "5th base"—5-methylcytosine; (2) DNA hydroxymethylation—further modification of 5-methylcytosine with appearance of the "6th base" 5-hydroxymethylcytosine; (3) various covalent modifications of nuclear proteins—histones (methylation, phosphorylation, ribosylation, ubiquitination, acetylation); (4) chromatin remodeling associated with special proteins; and (5) microRNAs and interfering RNAs effects. Some of these processes are comparatively well studied (DNA methylation, histones modifications), while others are still only identified and wait to be characterized in details. It is not always clear to what extent molecular mechanisms that are well identified in different species are relevant for humans, but in general existing and accumulating data give the picture of more or less universal mechanisms (Fig. 3.1).

DNA methylation. The existence of a "minor base," methylcytosine, in calf DNA became known after the publication of Hotchkiss (1948),

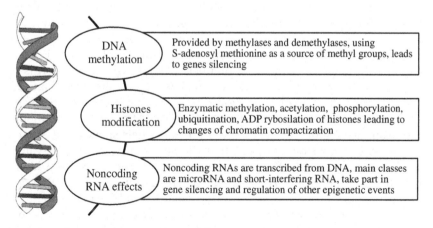

FIGURE 3.1   Main molecular mechanisms of epigenetics.

20 years after that Griffith and Mahler (1969) suggested that methylation of cytosine may be a mechanism of genes expression regulation. Almost at the same time research group of Boris Vanyushin started studies in this field (Vanyushin, Tkacheva, & Belozersky, 1970). For many years, this group systematically proved the significance of DNA methylation for regulation of genes. These researchers demonstrated the roles of this phenomenon in aging, stress, memory formation, and cell differentiation (Vanyushin, 2006). It is known that methylation happens to cytosine in those sites, where the latter precedes guanine, the so-called CpG pairs, which are palindromes—sequences that could be read in both direction on the complimentary strands of DNA. In human nonembryonic cells, methylated sites are distributed globally around genome and approximately 80% of CpG in DNA are methylated. Unmethylated "islands" of CpGs are mostly associated with promoter regions of genes (Jaenisch & Bird, 2003). Methylation/demethylation of these islands is extremely important—it results in the critical gene transcription changes due to changed pattern of reactivity to different binding proteins that are blocking genes activity. Methylation is provided by special enzymes, cytosine-DNA methyltransferases, which change their activity in the course of the cell life cycle. S-adenosyl methionine and maybe other methyl-containing substrates like betaine or folic acid may serve as donors of methyl groups. This fact makes it clear that methylation of DNA may be dependent on nutritional factors, and as will be seen further, it is one of the most well-studied and potent mechanism that impacts genes programming during development and influences health. The pattern of methylation is specific for different tissues (Szyf, 2011). Probably, this is a key mechanism due to which cells of different tissues possess specific sets of the transcribed genes (Richards, 2006).

Methylation of CpG islands in the promoter regions of genes functionally is usually associated with genes silencing. This may be due to methyl-CpG binding proteins which recruit proteins to the promoter of the gene, thereby blocking transcription or due to more direct blocking of specific DNA sequences recognition by transcription factors (Tammen, Friso, & Choi, 2013; Dupont, Armant, & Brenner, 2009). DNA methylation is reversible; removal of methyl labels is realized by different mechanisms, possibly with the help of reparation systems or due to the activity of special enzymes—DNA demethylases (Szyf, 2011). Although methylation of the promoter regions of genes usually means their silencing, it should, however, be taken into account that the situation is not single-valued in all cases; a number of "silent" demethylated genes have been identified. Besides, methylated/demethylated cytosines in intergene DNA may have dynamics which is different from "islands" and the role of methylation pattern outside coding sequences is not yet understood (Jaenisch & Bird, 2003). Transmission of the methyl labels to subsequent generations of cells during DNA replication is provided with the help of the so-called maintenance DNA methyltransferase. This enzyme recognizes semimethylated CpG regions and binds a methyl group to the nonmethylated cytosine in the newly synthetized DNA strand. In such a way this mechanism uses preexisting methyl labels for their reproduction and ensures that programmed DNA methylation pattern remains stable and goes further through cellular generations. Other DNA methylases (called de novo methylases) establish this pattern during development and provide for maternal imprinting of genes in oocytes. De novo methylation may also occur in differentiated cells, though at very low level. Methylation is regulated on many factors including genetic code itself (Tammen et al., 2013). Thus, epigenetics depends on genetics, while genes transcription is regulated by epigenetics.

5-Methylcytosine hydroxylation. Quite recent studies suggest that 5-hydroxymethycytosine—a product of hydroxylation of 5-methylcytosine by specific enzyme called TET—may also be an important component of the epigenetic signals dynamics. It is known that in mammalian DNA a very small amount of cytosines are hydroxylated. Interestingly, in the nervous system 5-hydroxymethycytosine is present in higher amount than in other tissues. The existence of this minor base may be the result of a series of metabolic transformation of methylcytosine leading to its demethylation. Hydroxymethylation of cytosines in the gene promoters leads to the reversal of gene silencing caused by methylation, so it may constitute one more contour of genes activity regulation (Tammen et al., 2013).

Histones modifications. Another key mechanism of epigenetic regulation is based on modification of the packing of DNA in the nucleosomes. Nucleosome is the basic unit of chromatin; it is composed of ~146 base pairs of DNA wrapped around an octamer of the four core histones (H2A, H2B, H3, and H4). The core histones are tightly packed in globular regions, while linker histone H1 interacts with DNA loop between nucleosomes. This organization of chromatin allows DNA to be tightly packaged, accurately replicated, and sorted into daughter cells during cellular division. In nondividing cells, chromatin may be present in two forms—euchromatin and heterochromatin. Euchromatin constitutes relaxed and more accessible state of DNA where transcription is more probable, while heterochromatin is a stable state, which prevents not only transcription, but also mutations and translocations. Heterochromatin is a prevalent chromatin state, while euchromatin is a temporal state directly associated with the time frame for transcription (Groth, Rocha, Verreault, & Almouzni, 2007).

Transition between two states of chromatine depends on posttranscriptional modifications of core histones. Terminal amino acid "tails" of molecules of histones protrude from the borders of dense nucleosome packages and serve as targets for a variety of covalent modifications. These modifications (involving lysine, arginine, and threonine residues of free tails) are mediated by an extensive set of the corresponding enzymes. By the nature of the modifying molecule, these processes include methylation, acetylation, ribosylation, ubiquitination, or sumoylation. The greater the level of saturation of histone monomers by methyl groups or more complex residues of acids, ribose, or small conservative protein ubiquitin, the lesser the level of DNA compactness, and the greater is the probability of transcription of certain regions of the genome. Relaxed state of chromatin allows not only active transcription, but also other DNA-related processes like recombination, replication, and DNA repair (Groth et al., 2007).

Enzymes providing covalent modifications of histones (acetyltranferases, phosphokinases, methylases, etc.) have antagonists providing reverse reactions (deacetylases, phosphatases, demethylases). Different modifications may compete with each other for functional groups of lysine and other amino acids. Most common modifications to histone tails are produced by histone acetyltransferases; it changes the ionic balance of histones and influences their binding to acid DNA, thus providing the most potent mechanism of transcription activation (Bernstein, Meissner, & Lander, 2007). In contrast, methylation of histones tails by histone methytransferase does not change the ionic balance; in this case change in chromatin activity is dependent on binding of proteins that serve as chromatin factors. The type of modifications is functionally significant, for example, methylation of histones inhibits transcription, while other modifications intensify this process (Bernstein et al., 2007). As a result, the probability for very dynamic regulation of DNA activity, both activation and silencing, short-time and long-lasting is ensured. Modifications of histones amino acid tails (for instance, lysine) "attract" the corresponding enzyme, which promotes modifications of new histone molecules. It may result in transmission of the protein labels in the process of DNA replication while recruiting new histones. DNA methylation and proteins modifications are functionally connected, for example, methylated regions of DNA attract histone deacetylases, which results in genes silencing. Conversely, histone modifications are involved in triggering de novo DNA methylation. Such interactions contribute to formation of multiple functional feedback loops, due to which some chromatin states may either reinforce or inhibit each other, thus providing stable and transmitted to further generation profiles of genes expression (Jaenisch & Bird, 2003). Therefore, all these interactions form a system of signals, which is capable of self-propagation and self-maintenance. The fact that DNA methylation and histones modifications are correlated with each other creates an important contour of the genome regulation. On the contrary, DNA methylation is supposed to produce more stable states, while histones modifications are more dynamic.

The most intriguing issue is how histone modifications can be transferred to subsequent cell divisions (or generations). It is generally known that during replication in eukaryotes each nucleosome is disrupted when "fork" passes and further reassembles due to physical–chemical interactions of DNA with proteins. For histones that act as core structure for DNA supercoiling, there must exist templating mechanism to establish corresponding modifications, otherwise in subsequent generations such marks will be "diluted" (Goldberg et al., 2007). These mechanisms remain largely obscure, though some data provide clues how it may happen. Parental histones bring their marks directly, and there is a theory that their marks may serve as blueprints for establishing the same profile on newly synthetized proteins (Groth et al., 2007). One of the possibilities is already mentioned recruitment of enzymes by existing modified sites.

In view of this it is interesting to remember that initially the idea of matrix synthesis, that is, the concept that one polymeric molecule may serve as a blueprint for synthesis of another copy, was developed by Nikolay Koltsov in relation to proteins, not DNA. Later Watson and Crick have explained mechanism of DNA replication which was so clear, simple, and exciting that possible role of proteins in propagating heritable information has escaped attention of specialists for decades. Now when 60 years have passed, complexity turns back and the whole mechanism of information transfer again becomes vague and indefinite. The situation is even more complicated so far as marked and unmarked histones should not only repeat modifications pattern but also occupy their genomic places, thus providing stable pattern of transcription in the cell lines. Most recent papers in this field are discussing several models of histones pattern transference each having some experimental support—the template model and assembly model (Ramachandran & Henikoff, 2015). Also, possibly due to existing interrelations between DNA methylation and histones modifications more stable DNA marks can somehow organize all further machinery of histones marks and location patterns (Groth et al., 2007).

Chromatin remodeling. There are reports that ATP-dependent and Polycomb proteins associated mechanisms of chromatin remodeling are also involved in epigenetic effects. ATP-dependent chromatin remodeling protein complex can bind to nucleosome and disconnect DNA from histones producing a transient DNA loop. Polycomb proteins are epigenetic regulators known to repress transcription by maintaining such state of heterochromatin, in which transcription factors are prevented from interaction with promoter regions of genes (Tammen et al., 2013).

Noncoding RNAs. Another interesting and potent mechanism is associated with nonencoding microRNA molecules. Findings in this field and especially elucidation of the mechanisms of RNA interference invite us to another world—world of regulatory RNAs. This parallel world was practically not known a couple of decades before. Functions of RNAs have always been associated with information transfer from DNA to ribosomes or ribosomes structure maintenance, but they appeared to be much wider. Regulatory role of noncoding RNAs seems to be even more important than those that we knew before. MicroRNAs are comparatively small (from 20 to 30 nucleotides of length) molecules that provide an independent way of genome fine-tuning for challenges of the environment. MicroRNAs are involved in the regulation of gene activity by direct inactivation of some genome regions via posttranscriptional mechanisms: they inhibit translation by binding to mRNA or activate RNAses (Tammen et al., 2013).

An interesting peculiarity of microRNAs is that this system is also self-reproducing and self-accelerating. Products of the activity of RNAses can serve as primers for synthesis of new microRNAs; this mechanism is related to methylation of DNA and modifications of histones so far as microRNAs are involved in regulation of the above-mentioned processes. All these facts give an impression how epigenetic mechanisms form internal cascades of regulation of the genome activity controlled by positive feedbacks. They seem to be able of self-reproduction and amplification of the signal, while self-supportive mechanisms may maintain their transmission in generations. On the contrary, a variety of the above-listed mechanisms of regulation of the activity of genes and their interaction (and counteraction) may be the reasons for removal of epigenetic labels; it may hypothesize that due to it the pattern of gene activity may be reversed, which explains fading of epigenetically induced phenotypes.

In general, existing knowledge about molecular basis of epigenetic events is in good concordance with the main features of epigenetic inheritance—the possibility of conservation of environmental influences by establishing epigenetic marcs and dependent phenotypes, often hidden, their realization of these phenotypes after sometimes long period of silence due to genes–environment interactions and new environmental cues. Epigenetic mechanisms also explains cases when environmentally induced phenotypes become heritable and pass through generations in a sex-specific manner, and cases when heritable traits disappear after several generations pass. All these events can be explained by the molecular machinery outlined here, though many details are still not well understood.

EPIGENETICS AND ONTOGENETIC PROGRAMMING

Every organism from biological point of view goes through life cycle that starts with egg fertilization and includes development, nutrition, socialization, reproduction, parenting, and aging. These basic events that actually constitute the life-span and give sense to existence are accompanied by environmental threats (stresses), rewarding impulses, changes in nutrition and diet, influences of physical and chemical environment, social interactions, and experiences accumulation. The whole life of the organism is associated with two main types of biological processes which have long-term consequences. One is heredity, that is, development and functioning in accordance with the inherited genetic program. Another is plasticity—the possibility to build different phenotypic variants on the basis of the inherited genotype. There are also numerous short-term adaptations (acclimatization) which are not taken into consideration so far as they do not fall under the concept of epigenetics. Developmental plasticity attracts much attention from the biological (evolution) and medical (health–disease) perspective. Development, evolution, and epigenetic mechanisms, when logically taken together ("evo-devo" synthesis), in application to child development open perspectives of deep understanding of the growth of the child, role of environmental influences, and changes of prevalence of some states and diseases in a comparatively short historical perspective (Hochberg, 2012).

From the point of view of biological success of the population or of a single organism, developmental plasticity is aimed to provide best chances for survival and reproduction, as well as for gaining better social position and passing important traits to next generations. Plasticity is the concept mostly congruent with epigenetics because it means building of different phenotypes on the basis of the same genotype. This is realized due to existence of epigenome—the whole set of chemical marks that make genes work in the most adaptive manner and establish conserved traits that must serve the goals of further adaptation. On the contrary, some phenotypes that are established during early phases of development may cause problems with physical and mental health in a long run, and this may become evident only later in life.

The epigenome provides long-term adaptation of the organism to the challenges of the changing environment. If we use the modern language of computer technologies, the epigenome serves as a dynamic interface between the very changeable environment and the relatively stable genome (Szyf, 2009a). Changes in the environment (environmental cues) are perceived as signals important for survival and reproduction; these signals "impose prints" on the genome starting from the very early stages of development (especially within the prenatal period, but also postnatally during further development) and shape the pattern of gene expression for the further life of the individual. This provides conformity of activities of biological systems to expected environmental conditions within the entire period of existence of the organism. This process is called ontogenetic programming. At the same time, if future environmental conditions in a broad ecological context do not match the expectations of the genome, the programmed functions can acquire a maladaptive pattern (Gluckman & Hanson, 2004; Hochberg et al., 2011).

From the general biological perspective, early stages of development prepare organism for all future life, and every stage of development has some predictive power regarding time frame and outcomes of each subsequent stage and further. Such views are considerably based on the evolutionary concept developed by a prominent zoologist and evolutionist, Ivan Ivanovich Schmalhausen, the author of the theory of stabilizing selection (Golubovskii, 2000; Hochberg et al., 2011). This can be seen in different species as signs of "being prepared" to most general ecological factors like seasonality, changing climate, availability of food, and existence of predators. The most widely cited examples of such correlation are, for instance, cases when newborn pups of voles conceived in autumn are born with thicker coat as compared with the same voles conceived in spring, or cases when some plant and animal organisms can acquire absolutely different morphological types (morphs) in response to environmental threats. These sometimes very profound differentiations emerge without any changes in DNA sequence but, as it becomes clear now, due to epigenetic mechanisms, which involve changing transcription pattern. These changes may appear in generations if the external factor persists, but may disappear or reverse in subsequent generations if ecological situation changes. The fact that they can be seen in several generations and then fade, make the whole picture of epigenetic inheritance blurred and contradictory to main rules of inheritance. In case of mammals (including humans), these mechanisms may be very important for health and disease, especially taking into consideration complex processes that happen during development and existence of several periods of development. As compared with other mammals human newborns have the most prolonged period of extrauterine development and maturation, including parental care, experience accumulation, and socialization. There is a rapidly growing body of evidence that epigenetic transformations are involved in all these components and that brain is the organ that is very much influenced by epigenetic marks.

During development human organism goes through several stages which are interconnected in such a way that every preceding stage, to a certain extent, predicts further qualities. For instance, everything starts with the egg fertilization and subsequent cell divisions and differentiation of zygote. The intrauterine life is divided into embryonic period when all structures of the body are developing, and fetal period when developed organs continue to grow and accumulate environmental influences (intrauterine environment). Perinatal period (around birth) covers last weeks of gestation and a week after gestation with the main transition stage—birth. Termination of the intrauterine life and exposure to external environment is one of the most important periods of rapid change, the way how the process of child delivery goes may itself have long-lasting consequences. Further biological preadult periodization includes infancy, childhood,

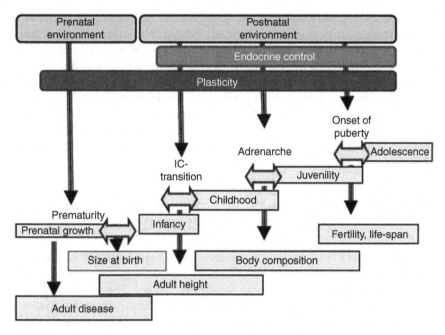

FIGURE 3.2 **Schematization of main transition periods during preadult human development.** *Source: Hochberg et al. (2011). Child health, developmental plasticity, and epigenetic programming. Endocrine Reviews, 32, 159–224.*

juvenility, and adolescence. The whole life of the individual (especially early periods) is the alternation of transition periods when development is fast and decisive and programs long-lasting effects (epigenesis), and more calm periods of growth and accumulation of body mass, skills, intellectual and social experiences, and so on (preformation). There is a growing body of evidence that environmental influences associated with epigenetic transformations that happen in transitional (critical or sensitive) periods may have crucial impact on further body forms, health, diseases, longevity, behavior, and adaptation abilities of the individual. Recently, Hochberg et al. (2011) after analyzing huge animal and human data have presented a scheme (Fig. 3.2) of biological periodization of preadult period of development of the human organism which considers transitions (turning point) between phases as periods of highest adaptive plasticity.

According to this model, prenatal growth predicts adult health and disease in the most general manner, transition from infancy to childhood preferentially determines adult height, changes from childhood to juvenility (adrenarche) have delayed influence on body composition, and transition from juvenility to adolescence (puberty) establishes longevity of reproductive activity and fecundity (Hochberg et al., 2011). It is interesting that Lysenko in his observations of plants was the first to differentiate between development and growth, and the idea that periods of transition (fast development) in humans are mostly susceptible to adaptive plasticity coincides with his views regarding plants. In his perception, inheritance of the organisms in such periods was "getting loose" which was the subject of great criticism from the point of view of classical genetics who could not forgive his deviations from gene-centered dogmatism (Zhivotovsky, 2014). Today, we have quite a different look at these ideas, and it will be possible to see that all predictive potential of early periods regarding further life trajectory is associated with epigenetic molecular events, which change the landscape without touching DNA sequence.

EPIGENETIC EVENTS IN PREEMBRYONIC AND EARLY EMBRYONIC DEVELOPMENT

Most massive and large-scale epigenetic transformations happen at the earliest stages of life—during gametogenesis, egg fertilization, and further embryonic development (Holliday, 2006). These changes are global because all epigenetic marks that are acquired during differentiation and across the life-span in parental organism should be erased in the germ-line cells to restore totipotency, so that new organism composed of two parental cells will start as "tabula rasa." Actually, preparation for sexual reproduction is a three-step process consisting of subsequent events: (1) erasure of somatic signatures in the germ cell precursors (the so-called primordial germ cells); (2) establishment of sex-specific and germ cell-specific epigenetic signatures and transcription profiles that enable processes of meiotic maturation and fertilization, and (3) the post-fertilization removal of these signatures to start the embryonic developmental program and to begin a new life cycle (Messerschmidt, Knowles, & Solter, 2014). This is associated with massive waves of methylation/demethylation, which must occur in a highly concerted manner to establish epigenomic state of important regulatory parts of genome like promoter regions, enhancers, and transposable elements. All this is crucial for normal development of the new organism; abnormalities in enzymatic machinery that secures these processes may result in serious disruptions in embryo development, including early death in utero and miscarriage (Messerschmidt et al., 2014).

Some epigenetic marks that come with germ-line cells from parents may cause several well-registered effects known as genomic imprinting or "parent of origin" effects (Pennisi, 2001; Richards, 2006). These DNA methylation marks appear during gametes differentiation and remain stable throughout life. Until now it is known that about 100 genes are imprinted in germ cells, effects from maternal and paternal sides are different and seem to be pursuing different goals from the point of view of development of child. If these influences are well balanced, genomic imprinting ensures successful development; if balance is severely impaired, it may cause several pathological syndromes, which are accompanied by development abnormalities, mental health retardation, and behavioral disturbances. For instance, DNA and histones methylation peculiarities may cause such human diseases as Prader–Willi and Angelman syndromes, both are associated with paternally and maternally imprinted genes inactivation on the 15th chromosome in the chromosomal region 15q11-13. The first one is characterized by overgrowth of the child (macrosomia), while in case of the second one children are much smaller than average for their age (microsomia). Other known conditions are Beckwith–Wiedemann and Silver–Russel syndromes, also with opposite phenotypes and associated with genes silencing at chromosome band 11p15. Both types of syndromes are accompanied with neurological abnormalities and behavioral peculiarities—in case of overgrowth Prader–Willi phenotype children are mostly retarded and autistic, while small-sized and underweight Angelman children have a very peculiar constellation of emotional and behavioral "happy puppets" symptoms. These abnormalities of different severity are linked to parental imprinting of genes in placenta and regulation of nutrients transfer from mother organism to a fetus, which affects body growth in the womb and behavior after birth (Reik & Walter, 2001).

Deep and very conservative biological mechanisms surrounding very early development and its epigenetic background are coming into conflict with modern technologies and trends in medical practice. So far as genomic imprinting and epigenetic reprogramming happen during gametogenesis and the most early preimplantation period, it raises the question if different environmental influences during the procedure of assisted reproductive technologies may interfere with maintenance of the imprints. By now millions of people have been born due to in vitro fertilization, and there is already enough data to have epidemiological results of possible health problems. Indeed, the occurrence of Beckwith–Wiedemann and Silver–Russel syndromes in this population is several times higher than in controls; some abnormal imprinted loci were identified in children conceived by in vitro fertilization (Le Bouc et al., 2010). There is also a vivid discussion if children conceived in vitro may have some mental health problems later in life.

Other events may be linked to basic biological mechanisms of child development. So far as genomic imprinting is a parent-of-origin effect, an elegant hypothesis has been formulated that "battle of the sexes" (competition between imprinted genes) actually forms two alternative strategies of body and brain programming, which have consequences not only for body weight, but also for personality traits and probability of psychiatric disorders. The latter include psychosis, which may appear if maternal influences are prevailing, and autism spectrum disorders, which is the result of paternal influences prevailing (Badcock & Crespi, 2008). It is hypothesized that paternal genes are aimed to invest in every newborn child by promoting body growth and autistic behavior, which demands a lot of maternal input in terms of nutrition and caregiving, while maternal genes are doing the opposite—promote lower birth weight and higher socialization traits to save resources for subsequent children and make a child more easy to handle (Badcock & Crespi, 2008). This hypothesis is supported by comparison of clinical features of Prader–Willi/Angelman and Beckwith–Wiedemann/Silver–Russel syndromes. It may also be linked to external conditions like food resources or hostile threats. For instance, alternative strategies may be involved—with less children but with bigger investment in a single child and stable relationship, and with many children and with lower investment in their development, socialization, maturation, and with frequently changing relationships, and so on. These strategies may be beneficial in different (rich and stable, or hungry and threatening and hostile) environments and may represent alternative life models. It is interesting in connection with this to recall again naïve genetic theory of Lysenko, who was expressing views that domination of maternal and paternal traits depends on the environmental needs and conditions (Zhivotovsky, 2014). One can imagine how such views were met by classical geneticists who have just built their harmonious system based on Mendelian inheritance, dominant and recessive alleles, and so on.

Another well-known effect linked to parental epigenetic processes is X-chromosome inactivation (compactization) in female mammals. It is a random process, which is supposed to happen at time of gastrulation in the epiblast (cells that will give rise to the embryo). The randomness of the pattern of inactivation (paternal or maternal) in each cell clone remains stable for the rest of the life. The aim of X-inactivation is thought to be a compensation of the genes dosage of this chromosome in the male and female organisms. It is also a remarkable example of epigenetic programming so far as approximately 1000 genes are silenced and packaged in the form of compacted chromatin (Barr bodies). The process of X-inactivation is a subject for plasticity and reprogramming—during oogenesis inactivated X-chromosome is reactivated, but later is again inactivated in next generation. The process of inactivation is associated with DNA methylation, histones modifications, and with effects on noncoding RNAs (Chow, Yen, Ziesche, & Brown, 2005). Although X-inactivation is a normative process, full loss of X-chromosome in females results in Turner syndrome. In this condition, which has both phenotypic (appearance) and cognitive symptoms, difference in imprinted genes in case paternal or maternal X-chromosome is lost is reported, which may have an impact on severity of cognitive impairment (Hochberg et al., 2011).

The phenomena described above give an impression how important are epigenetic effects and how deeply they are involved in the transfer of some phenotypes from parents to offspring without touching DNA sequence. Such processes like genomic imprinting and X-inactivation are important for early stages of development and growth and their deregulation may cause serious damage for the developing organism. Even subtle influences that cause changes in the normative genomic imprinting may be deleterious. The same type of events may happen also at later stages of development and they may their own consequences for later life. They are associated with such important factors as nutrition, stress, and experience accumulation.

NUTRITION, BODY GROWTH, AND EPIGENETICS OF OLDER AGE DISEASES

Since most early stages of embryogenesis are passed properly and differentiations have established normative structure of the body, organs, and tissues, further main goals of the intrauterine organism are growing and accumulating of the body weight until the birth will start a new stage of development. One of the most probable natural situations is undernutrition in pregnancy—that is, calories and important nutrients supply restriction for the embryo and fetus. It is also of great importance for the newborn shortly after birth so far the neonate is fully dependent on mother care and nutrition. It may be said that this situation means the first serious stress that is endured by the organism in its life. Hunger is a serious stress for a pregnant female, and embryo or fetus in such situation is experiencing mixed influences—both from nutrients shortage and from cortisol excess so far as mother experiences severe stress.

There are many data today confirming that epigenetic regulation is very important for human in utero growth and that timing and severity of in utero calories restriction has long-lasting consequences later in life. For instance, early life undernutrition and stress enhance adult risk of developing metabolic diseases like type 2 diabetes and cardiovascular pathology. Epidemiological studies have shown that individuals, who were born small-for-gestational age, have higher risk of cardiovascular morbidity and mortality when they reach adult age. So, these age-related diseases can be considered to be at least in part a prenatal or early age (pediatric) diseases (Hochberg et al., 2011).

In relation to the above-described data, the hypothesis of the "thrifty phenotype" attracts attention. This hypothesis is based on observations that abnormally low body mass of the child at birth is associated with the development of type 2 diabetes because of permanent disorders in the glucose–insulin system, disturbances of appetite regulation, and abnormal function of pancreatic β-cells. The formation of the "thrifty phenotype" can be interpreted as a consequence of the "semistarvation" status of the fetus within the intrauterine period. It induces a compensating behavioral style within the conscious age, which is directed toward constant recovery and storage of resources (Hales & Barker, 2001). From this point of view, obesity epidemics among teenagers and the prevalence of the metabolic syndrome among middle-aged persons in developed countries in recent decades can be viewed as a reflection of the widespread availability of high-calories food for generations, whose parents had no possibilities for adequate nutrition in their young years and/or were subjected to stress within the period of maturation and transition to the reproductive age. Thus, hungry childhood or adolescence of parents may program obesity in later life of their children. This is a very good example how social factors (society development, rise in the living standards, better nutrition, plenty of cheap and attractive fast-food, etc.) may interact with deeply embedded biological mechanisms. There will be many opportunities to see similar interactions during further discussion on integrative mechanisms that put together social, interpersonal, and biological factors associated with stress.

If early stages of body growth of a child were restricted, active "spurt" in body weight accumulation after birth may result in enhanced central (belly) fat accumulation in older age, which appeared a risk factor for early cardiac infarction. Many of these effects are thought to be associated with abnormal IGF2 gene methylation, the gene responsible for insulin-like growth factor, a peptide that has insulin-like regulatory activity and enhances body growth (Hochberg et al., 2011). Human history knows a lot of situations when big contingents of people were starving, so much attention was paid to epidemiological studies that provide such quasiexperimental conditions. Probably the most well-known study is dedicated to survivors of the so-called "Dutch famine" of "Hongerwinter"—a famine in the north of Netherlands in the winter of 1944–45, when approximately 4.5 million people remained in a blockade without food supply. It was found that individuals who were prenatally exposed to famine in 1944–45, 6 decades later, in the beginning of the new century, had much more morbidity and mortality associated with a variety of metabolic and cardiovascular disruptions like hyperlipidemia, type 2 diabetes, and hypertension. Later it was revealed that less DNA methylation of the imprinted IGF2 gene was found in exposed survivors as compared with their nonexposed siblings (Vaiserman, 2015b). Similar results have been obtained in other human population studies, for instance, a study from West Africa reports that in children conceived during the peak rainy season months, when food is less available to their mothers, levels of DNA methylation were higher at regions of genes associated with such diseases as Tourette's syndrome and hypothyroidism (Tokunaga, Takahashi, De Meester, & Wilson, 2013).

The studies with famine cohorts have confirmed that periconceptional period is more vulnerable as compared with late gestational period, indicating once more that most early life stages are extremely vulnerable to nutritional restrictions (Vaiserman, 2015a). Hunger during pregnancy can be easily modeled using rodents. Epigenetic changes found in experimental studies in offspring of starved pregnant rats and mice can persist in several generations even if food is available. On the contrary, methyl groups' supplementation may alter outcomes. Thus, lowered proteins intake during pregnancy in mice leads to lower methylation of genes involved in carbohydrates metabolism in different organs of offspring in several generations, while folic acid supplementation prevented this effect (Lollicrop, Phillips, Jackson, Hanson, & Burdge, 2005). These data testify on the possibility of transgenerational transmission of poor nutrition effects and their reversibility by diet.

It is not only starving but also obesity and high-fat and sugars diet (analog of a modern westernized food) may be the reasons for epigenetic programming in further generations. There are reports from animal studies that offspring of male overweight rats mated with normal female rats exhibit impaired glucose tolerance and insulin secretion as adults. Similar data are found in human populations. Overweight fathers can program epigenetic changes in their daughters—an effect transmitted by their sperm and induced by consumption of high-fat diet (Tokunaga et al., 2013). There are data that availability of food at certain sensitive periods (for instance, slow growth period from 8 to 12 years, just before puberty) has long-lasting sex-specific transgenerational consequences (Tokunaga et al., 2013). Overweight mothers are also producing negative effects on their children weight and transfer propensity to metabolic syndrome through abnormal intrauterine environment. Thus, increasing prevalence of excessive body weight and obesity due to imbalanced nutrition forms a vicious cycle leading to similar or related abnormalities in the progeny (Hochberg et al., 2011).

Not only calories but also essential nutrients and nutraceuticals are able to influence epigenetic processes. Sources of methyl group are mostly important, which is quite easy to understand. Possibly the most well-known data are about the possibility to change effect of agouti gene in mice. Mice that are heterozygous for the agouti yellow allele have yellow coats and tendencies toward obesity, diabetes, and cancer. Supplementation of pregnant mouse diet with methyl donor S-adenosyl methionine leads to methylation of the agouti gene in fetuses and leads to birth of brown-coated mice, which are free from above-mentioned risks. These mice remain healthy all their lives and have no signs of diabetes—an example of maternally heritable epigenetic-based phenotype (Wolff, Kodell, Moore, & Cooney, 1998). Other bioactive food ingredients can affect the methylation status of genes in different ways, for instance, biologically active polyphenols from green tea reduce DNA methylation in cancer cells through the competitive inhibition of DNA methyltransferase (Tammen et al., 2013). This finding is often mentioned in view of reduced risk of cancer in oriental peoples (China, Japan, Sri Lanka) who consume a lot of green tea. Many studies are devoted to soy isoflavones (phytoestrogens) known for their ability to lower cancer risk and influence reproductive system of females. Supplementation of these natural food components to experimental animals results in hypermethylation and silencing of certain proto-oncogenes. Methylation status of oncosupressors was also changed. This can be one of the explanations of lowered risk of cancer in Asia populations where soy is more widely used in diet as compared with Europeans (Dolinoy, Weidman, & Jirtle, 2007). The most impressive evidence comes from China—a group studying miRNA in humans and plants have revealed that rice miRNA can enter blood and serum of humans through food intake and can influence genes in the liver that are involved in low-density lipoproteins levels regulation. It also suggests explanation why several diseases show lower prevalence in these populations (Zhang et al., 2011). It has been known for centuries that diet can influence health, but recent data provide quite different understanding how this influence may be actually realized.

Recently, it became evident that quite many nutrients have capabilities to modify epigenome. Among them are microelements, vitamins, polyphenols, saturated and unsaturated fatty acids, polysaccharides, aminoacids, carotenoids, and other food components (Tammen et al., 2013; Tokunaga et al., 2013; Kanherkar, Bhatia-Dey, & Csoka, 2014). Their effects can be seen both in programming effect of pregnant maternal diet on offspring health and in early life programming heath and disease of the individual in older life. Elucidation of their role in health promotion, cancer prophylactics, disease susceptibility, and longevity changes our perception of the role of well-known nutrients, including vitamins, which were traditionally seen only as coenzymes predecessors and regulators of metabolism. In relation to all these facts, it is reasonable to recollect an oriental concept of health, which in contrast to western concept (genetics, life-style, environmental factors, and level of medical aid) says that our health depends on what we eat (30%) and what we think (70%). Modern data on epigenetic programming by nutrients strongly support an ancient paradigm "we are what we eat." It is also appropriate to cite here often mentioned opinion that populations occupying certain territories for a long historical time in a natural way get accustomed to local nutrition, and this may have consequences from the point of view of national character, appearance, behaviors, and most common diseases. Therefore, it would be very interesting to understand to what extent epigenetics is involved in the second part of the well-known oriental wisdom—"we are what we think."

EPIGENETICS OF ECOLOGICAL HAZARDS, IN AGING AND CANCER

Each individual across the life-span is surrounded by many ecological hazards, and in the recent time their role is obviously growing. These exposures are often referred as ecological stress, one of the components of the modern global stress. For decades, negative effects of natural toxins, microelements, and new anthropogenic pollutants (xenobiotics) were usually understood as the result of accumulating influences that damage certain biochemical pathways or cause oxidative stress and ultimately contribute to higher morbidity and mortality, including risk of cancer. It is generally acknowledged that organism in its early life periods is mostly susceptible to environmental chemicals and hazardous physical factors like ionizing or nonionizing irradiation. Recent data provide exciting evidence that chemical environment may not only have long-lasting effects after early exposures, but also can lead to transgenerational effects that can be traced in next generations in experimental and natural conditions even if exposure is not applied any more.

Although some microelements (selenium, zinc, copper, manganese, molybdenum, and cobalt) produce beneficial effects suppressing expression of harmful genes, others like lead, mercury, arsenic, and cadmium are causing adverse effects (Tokunaga et al., 2013; Kanherkar et al., 2014). Heavy metals which were known mostly as modifiers of peptides conformation thus causing their toxic effects on enzymes, as it was revealed recently involving all known epigenetic mechanisms: DNA methylation, histones modification, and chromatin remodeling and mostly interesting miRNA profile changes. This partly explains their in utero toxic effects leading to long-lasting disturbances like neurological or cardiovascular diseases and cancer (Vaiserman, Voitenko, & Mekhova, 2011). Interesting results are obtained regarding influence of the so-called endocrine disruptors—a class of pesticides that are structurally close to animal sex hormones. It appeared that many previously registered toxic effects of these compounds during pregnancy are due to epigenetic programming of embryo cells and tissues. Thus, intrauterine exposure to bisphenol leads to functional disturbances of reproductive system and to infertility in later life in females and high risk of prostate cancer in males. This is associated with changed profile in several genes methylation (Dolinoy et al., 2007). A very remarkable study aimed on evaluation of effects of antiandrogen agent vinclozaline has revealed that intrauterine exposure to this fungicide causes long-lasting consequences that can be traced in four generations. Male embryo exposed to vinclozaline obtained spermatogenic defect that was transmitted further in male line. The defect is associated with DNA methylation changes in germ line and affects not only spermatogenesis but also other developing tissues and cell types through the paternal epigenome (Skinner, Amway, Savenkova, Gore, & Crews, 2008). This is a very clear example how exposure of developing organism during the critical period to ecotoxicants can reprogram germ cells and how adverse effects can persist in several generations. It is very important that in this particular series of studies with vinclozaline it was conclusively proven that aberrations in spermatogenesis could be traced in several generations of descendants on the paternal line while exposure was not applied any longer and (what is important) there was no remaining toxicant in the body of descendants, otherwise these results might have been explained by persistent intoxication. It convincingly testifies that germ-line cells were specifically programmed and kept epigenetic marks, which were not erased (or somehow escaped) reprogramming during gametogenesis.

Another important issue is epigenetic effects of substances of abuse which are so widely spread in human population. These "social medications" play important role in the society, especially when psychosocial stress is enhanced. Their intake is, to a great extent, dictated by cultural peculiarities of the given society and is usually associated with life adversities, lower life standards, and poor socioeconomic status. Alcohol consumption on the population level shows an association with inequities, poverty, and distress. Alcohol, nicotine, and drugs of abuse consumption are growing in different social strata in situations of psychosocial stress (Rozanov, 2013a,b). Recent studies of drugs of abuse present a growing body of evidence that many known adverse effects of these chemicals may also be the result of epigenetic programming in brain and other tissues (Vaiserman, 2013). Alcohol has, for many years, been known as teratogen; paternal alcohol abuse is known to produce aberrant birth weight, heart defects, and cognitive deficits. In utero and in early life alcohol exposure produces many neurological, cognitive, and behavioral abnormalities. Very similar deficits are produced by in utero exposure to heroin, cocaine, and amphetamines and, to a certain extent, by smoking nicotine and cannabis. In case of amphetamines, higher risk of malignant neoplasms in the later life should be added. There is a significant body of evidence that these later-in-life abnormalities and health consequences are associated with active participation of all known epigenetic mechanisms—DNA methylation, histones modification, and miRNA effects. Recent reviews of studies in this field provide conclusive information that most widely involved neurochemical systems that are responsible for these delayed neurological and cognitive abnormalities are neurotrophins, HPA system components, IGF, protein kinase C, and p53 tumor suppressor peptide (Vaiserman, 2013).

Very impressive results are obtained regarding delayed health effects of cigarette smoking in humans. In one study it was found that smoking induces specific change in the spermatozoal microRNA content, in particular almost 30 miRNAs were expressed differently in smokers as compared with nonsmokers. These changes are thought to trigger apoptosis of cells and other abnormalities of sperm development, which may explain how harmful phenotypes can be induced in the progeny of smokers (Marczylo, Amoako, Konje, Gant, & Marczylo, 2012). It is remarkable that in such studies patrilineal transmission was observed and fathers were early smokers (preadolescent or early adolescence). Although in utero epigenetic programing by mother organism through placenta is better understood, paternal line of inheritance is not so clear. The only logical explanation is that environmental factors (smoke components) can reach spermatogenic tissue and produce epigenetic marks there. Findings of epigenetic programming of the sperm provide a very important line of evidence confirming this assumption. Smoking, alcohol, and other substances of abuse usually go together in males and often are seen as "sins of fathers." In view of this it is interesting to recollect views of some old medical doctors who, while discussing reasons of hepatitis in young patients, used the formula "father has spent on drinking liver of his son."

These data, together with new mechanisms of action of nutrients and ecological toxicants, give an impression how chemical environment can program health and disease of big populations. So far as ecological situation does not improve, on the contrary, pollution of water, air, and soil, as well as food products, is progressing, concerns about possible long-term consequences for health in new generations are growing. The same can be said about substances of abuse. It also demonstrates once more how epigenetics mediates the relationship between the genome and the environment; in this case chemical environment.

There is also accumulating data on epigenetic effects with multiple genes activation by such environmental pollutants as physical factors, including low-dosage ionizing irradiation (Ma, Liu, Jiao, Yang, & Liu, 2010) and, possibly, nonionizing low energy electromagnetic fields (Liu et al., 2015). From these point of view, recent findings about germ-line genomic instability that can persist in several generations after exposure of male animals to irradiation is a logical continuation of this line of evidence (Barber, Plumb, Boulton, Roux, & Dubrova, 2002).

Ecological toxicants, irradiations, and substances of abuse may be the reason of early aging. On the contrary, aging per se is the process which greatly involves epigenetic events. Early studies of Russian authors Grigoriy Berdyshev and Boris Vanyushin have revealed a global decrease of 5-methycytosine levels in aged fish and rats and proposed that this demethylation may have regulatory importance for the aging process (Hochberg et al., 2011). Later studies of human lymphocytes obtained from different age subjects from embryo to old age have revealed that some loci in the genome are hypomethylated, while other are hypermethylated, making the picture more complicated (Vaiserman et al., 2011). In some studies age-related epigenetic profiles were extensively studied in monozygotic twins. Monozygotic twins, though being genetically equal, often manifest differences in body weight, diseases susceptibility, as well as in psychological and behavioral profile. Well-known study of Fraga et al. (2005) has revealed that elderly monozygotic twins who spent their lives apart from their parental families had numerous epigenetic differences. These differences were more profound in older age pairs as compared with younger ones, and also were bigger in twin pairs who were living separately in comparison with those who were living in their parents' households (Fraga et al., 2005). This pairs-based study clearly indicates that epigenetic marks are changing during the life-span and that external environment (social, chemical, physical, etc.) may have either unifying or differentiating influence on epigenome. Later this conclusion was confirmed in longitudinal studies where twin pairs' epigenomes were assessed subsequently after a substantial period of time, from 11 to 16 years (Hochberg et al., 2011). From studies on epigenetics of aging, an important conclusion may be drawn—epigenetic events continue to accumulate during the whole life of the individual. Not only early periods of life, but also further periods of the life are susceptible to epigenetic modifications, which may occur in different organs and tissues of the organism and which may have importance for each further period of life. This is especially relevant for experiences accumulation and mental disorders, which will be discussed in future chapters.

Summing up, epigenetics is not a single process or event; it is based on different molecular processes. It is a universal mechanism that plays important role in the biology of all living objects, from plants to humans. The word "epigenetics" implies several distinct processes, but the essence of it is that modifications of chromatin once happened cause long-lasting effect that can be noticed phenotypically, often not immediately, but after a certain period of time, and may be reversible. This makes epigenetic events not very noticeable. It is sometimes not easy to couple of events that are divided by several decades and causal links between which are not evident. Who could say 10–15 year ago that most prevalent diseases that threaten modern human populations (cardiac failure, diabetes, cancer, mental disorders, etc.) have very early origin and are programmed by maternal or paternal influences other than DNA sequence? In the medical–biological paradigm, these conditions were understood as multifactorial, based on genetic predispositions which interact with environmental (mostly caused by life-styles) influences. Thus, development of a pathological condition was seen as interplay between genes and environment, or their interaction, and the late emergence of these "diseases of civilization" was explained as the result of long-drawn-out action of external factors and accumulation of morphological, physiological, and biochemical impairments. Revealing of the programmed developmental origin of these diseases and involvement of epigenetic events has changed the paradigm. The same is with specific nutrients, ecological toxicants, and substances of abuse—at least part of all health problems that usually appear after exposure to these components are due to very early epigenetic events and programmed genes expression.

Environmentally driven changes in epigenome may ensure not only long-lasting programming in a particular individual, but also can lead to germ cells programming and to transmission of the phenotype in several generations. It is important that mechanisms leading to such programming are sex-specific and have different time frame in males and females. All eggs are present in the female organism during prenatal period, while sperm maturation begins in males only with puberty. Consequently, females are bearing all experiences by maternal line in a very conservative way, while males are accumulating experiences (and epigenetic marks) due to exposures and behavior in the preadolescence or early adolescence period. Thus, not only in utero or early postnatal adverse influences mediated by maternal organism can have long-lasting consequences, but also time-framed influences in males that depict current ecological, behavioral and psychological situation, may be conserved and transmitted. All these facts and findings arise new thinking not only regarding health and disease development, but also regarding evolutionary processes, both of large time scale and in shorter periods of time.

EPIGENETIC PARADIGM AND EVOLUTIONARY BIOLOGY

"Nothing in biology makes sense except in the light of evolution" as it was said by Theodosius Grigoryevich Dobzhansky, one of the cohort of evolutionary biologists who was adjusting Darwin's views to the new knowledge regarding DNA structure, mechanisms of replication, and mutations. This phrase, which appeared in the 70th of the last century, remains mostly modern today, especially with regard to new epigenetic paradigm that has a great influence on the evolutionary theories. Epigenetic phenomena remained for a while very underestimated in respect of their value for adaptation of species and individual organisms to environment conditions. Classical genetics accustomed us many years to the idea that properties of an organism, all its phenotypes (morphological, metabolic and behavioral), are rather firmly determined by its genetic basis—a unique set of the genes received from ancestors. As extreme manifestation of such thinking can serve the well-known phrase of the father of DNA J. Watson—"We are our genes." This thinking (genetic determinism) was dominating even in such sphere as behavioral genetics, which studies differential input of genetic and environmental influences in development of psychological traits, behaviors, and mental disturbances (Holden, 2008).

The second key moment of "genes-centered thinking" is that genetic material is conservative, while mutations are rare and stochastic. Conservatism of genetic material was considered a pledge of stability of the species, while mutations together with ecological changes served an explanation of possibility of speciation. These two theoretical positions, which were dominating for a long period of time, served the basis of neo-Darwinism or modern evolutionary synthesis. The central idea of this theoretical system is natural selection and gradual and very slow evolution. Mutations serve as the "engine" of evolution, while millions of years may past until a chance will come and certain mutation will become an adaptive trait in a changing environment. As many of contemporary authors notice, this system of views was not paying enough attention to behavior and stress, which are the basic events of life of any organism. Moreover, development of the individual organisms was not perceived as the important component of the evolution of the species (Golubovskii, 2000; Gluckman & Hanson, 2004; Richards, 2006; Hochberg et al., 2011).

Emergence of epigenetics and epigenomics in its modern understanding has changed the paradigm by focusing attention more on environment than on genes. Epigenome is seen as dynamic interface between more or less stable genome and constantly changing environment. Great support for such thinking was provided by more detailed description of molecular mechanisms that take part in transcription regulation and their modulation by stressful factors of the environment. However, mechanisms by which environmental cues are actually imposing epigenetic marks are not known. There may be multiple ways, starting from availability of methyl donors and regulation of DNA methylation and histones modification machinery by metabolites, hormones, and other biologically active molecules to proteins and microRNAs interactions, and so on. It is quite possible that stress hormones are important regulators of epigenetic machinery. Another possibility is that epigenetic signatures in brain may be associated with electric impulses in the neurons that are reacting to stressful stimuli. Molecular pathways of establishment of epigenetic marks will be of course found some day, by now it is obvious that the variety of mechanisms that may lead to changes in epigenetic landscape is much wider than that for changes in DNA sequences. There are also several distinct and very basic differences between classical mutations and epigenetics transformations (Table 3.2). In the table, results of our analysis and some items from the publication of Burggren (2016) are used.

From comparison above, one can assume that epigenetic marks must be much more important from the point of view of effects of environmental cues on development, functions, and behaviors. It overcomes one of the shortcomings of the gene-centered thinking, which did not encounter behavioral and stress-related events. These events are mostly important and are inherent to life of any species, and the fact that we start to understand how they can program development, health, disease, and behavior

TABLE 3.2   Differences Between Mutations and Epigenetic Marks

	Mutations	Epigenetic marks
1	Stochastic event	Environment-driven event
2	May appear in any part of the genome, most probable in noncoding areas	Appear in regulatory regions, promotor or enhancer parts of genes
3	Rather rare event	Quite frequent, even constant event
4	Affect one or few individuals with subsequent slow increase (if advantageous) or decrease of phenotype (if deleterious)	Affects many individuals simultaneously, all receiving the same mark and similar phenotype
5	Permanently stay in the population until eliminated or reversed by new mutation	Transient but may be sustained if the environmental exposure is pertaining
6	Induced by physical and chemical environmental mutagenic factors	Induced by very different external influences and stresses, including social factors

is of crucial importance. Humans seem to be the most vulnerable creatures in this sense so far as their stress is linked to social processes, which are changing too fast and leave much less space for recreation and rest if compared with nonhumans.

In these environment-driven processes, there are two main biological mechanisms—context-dependent and germ-line-dependent epigenetic inheritance (Crews, 2008; Jablonka & Raz, 2009; Hochberg et al., 2011). In the first variant epigenetic marks and phenotypes that depend on them are transferred to next generations by behavioral mechanisms (early life stress and stress-related system programming, which will be discussed later) or due to persisting environmental stimulus or factor, like ecotoxicants exposure—the case, which has been considered in details earlier. In the second variant we are dealing with situations, when epigenetic marks have reached germ line and have escaped reprogramming during early embryonic stages of development. Mechanistically, it may be the result of transmission of noncoding RNAs and chromatin states within germ-line cells (Mandizabal, Keller, Zeng, & Soojin, 2014). In context-dependent epigenetic events, we observe transmission within a generation of cells or within an individual's own life-span, while in germ-line-dependent epigenetic events we are dealing with a transmission across generations. On the contrary, all depends on the environment factors. For instance, persistence of a stressor, or toxicant, or substance of abuse or behavior has a critical importance both in the context-dependent model and in the germ-line-dependent model. It may be theorized that in context-dependent model, if the environment or behavior changes, erasure of epigenetic marks and fading of the phenotype may happen earlier, in two to three generations, while in germ-line-dependent model it will need longer time and more generations.

Epigenetic programming and soft inheritance are shifting our understanding of evolution from neo-Darwinian approach, based on the role of long-lasting gradual changes, to acknowledging the role of comparatively fast transformations based on adaptive responses to changes in the external physical, chemical, or social environment. It is most likely that these two processes exist as parallel tendencies, supporting and complementing one another. It is interesting that understanding the role of epigenetic programming turns us back to the theory of Lamarck, whose central ideas were that variability arises under direct or indirect influence of the environment and is followed by adaptation, and that signs acquired in individual life are capable to be transferred to the next generations (are inheritable). Neo-Darwinian theory in some sense separates organism from the environment; it concentrates on stochastic mutations in the organism, while environment serves as a framework in which elimination or survival with better reproduction may occur. In contrast, Lamarck's theory is based on the idea that transformation of the organism is the result of the immediate experience from the environment. In gene-centered thinking, evolution is directed from genes to environment, while in epigenetic thinking it is directed from environment to genes.

Moreover, individual development of the organism is taken into consideration. In accordance with epigenetics, evolution is a tandem process involving first development with its periodization and flexible responses to environment, and then selection, which determines what variants will be supported, spread, and maintained (Crews, 2008). From this point of view it may be said that "environment talks to genes" (Szyf, 2013) and that "genes learn from environment" (Jaenisch & Bird, 2003). If we take individual behavior into consideration, it may be also said that "each individual is producing his/her own epigenetic landscape" so far as behavioral pattern may dramatically change the level of exposure to environmental stimuli. It draws us closer to ideas of such evolutionary biologists as Richard Goldschmidt, Ivan Schmalgauzen and, paradoxically, Trofim Lysenko with his insights and "feeling of living organisms" (Zhivotovsky, 2014).

Germ-line transmission of epigenetic marks and phenotypes and all philosophy of epigenetics as soft inheritance means serious blow to the "germ-plasm" theory of August Weissman and actually supports legacy of Lamarckism in its pure expression. Psychologically it is rather difficult to accept, but many modern authors quite clearly state in their publications that Lamarck was right, he was simply born in a wrong time (Richards, 2006; Crews, 2008). It is also important to remember that another central idea of Lamarck was that during adaptation organisms were "striving for excellence." Later followers of Lamarck have developed his view and introduced "psycholamarckism" which supported ideas that willed efforts of animals, their high mental activity, and memory are actually the main engine of evolution (Nazarov, 2007). Psycholamarckism may easily turn into vitalism if the psychic component of Lamarck theory is turned into an absolute. Although our materialistic mind refuses to perceive these ideas seriously, it is remarkable how many very professional physiologists and philosophers have seriously discussed these issues within the context of evolutionary thinking (Nazarov, 2007).

Novel data explain how information flow from environment to the epigenome shapes our bodies, and determine (at least to a certain extent) our behavior, health, and diseases. It gives plausible explanations of many structural and functional changes that can be seen in human population recently. For instance, European men became taller 13 cm for the last 150 years. The age of menarche in Western countries has decreased for the same period of time for 4 years. That means that in six generations biological changes have reached great extent, which is difficult (or even impossible) to explain by changes in DNA sequence or mutations (Hochberg et al., 2011). It may be added that there are sporadic reports about even more interesting facts—forehead height in contemporary people is statistically higher than in sculls that are 300–500 years old. An interesting study of an isolated population shows that microevolution can be detectable over relatively few generations in humans (Milot et al., 2011). All these fast changes cannot be explained by classical genetics and Darwinian selection, while epigenetics with transgenerational transmission provides a very logical explanation. It is possible that Darwinian and Lamarckian mechanisms are alternating in time and complement each other in relation to environmental change. Damiani (2007) has put forward an interesting idea that in the catabolic-entropic phase of development the genetic apparatus and Darwinian processes support selection and conservation of biological systems, while in the anabolic-syntropic phase epigenetic apparatus and the Lamarckian anti-Darwinian processes produce innovative adaptations. In the latter case, environmental stress plays the role of driver of the evolution (Damiani, 2007). The most intriguing is the probability that stress via epigenetics may be responsible for growing mental health problems and suicide seen in last decades and could be one of the signs of such processes. This issue is the main subject of our discussion and will be addressed in further chapters.

Epigenetics is a quickly developing system of views, a new paradigm in biological and medical science. It has a tendency to explain a lot of events, medical and biological, in more or less clear terms. However, a very professional view of the evolutionary biologist reveals many internal problems and inevitably leads to great number of questions which until now have no distinct answers. In spite of thousands of very enthusiastic publications, one of the recent reviews still states that "unfortunately, progress in epigenetic research is being hampered by a lack of clarity as to what is epigenetics" (Burggren, 2016). As every new and very quickly developing area of science, epigenetics may produce false results or suggest simple but maybe false conclusions. We consider that it is very important to integrate new knowledge with what we know by the moment and be careful with conclusions and maybe ask more questions, especially when we deal with such subject as human being and discuss such human issues as suicide.

It is now appropriate to turn back to what we started to discuss in this chapter—why Ivan Petrovich Pavlov and his colleagues could not find confirmation that conditioned reflexes may be transmitted in generations? It is possible to speculate that the reason was that in his studies only positive reinforcement (food) was used, while negative reinforcement (stress) may produce much stronger effect and can involve much more potent mechanisms of transmission of external signals to genes. Positive reinforcement like food is important for immediate behavioral result, but negative signals (hunger, starvation, pain, fear, frustration) are much more important for survival and long-lasting results. Most of the recent studies, which prove the role of epigenetics in memory formation, especially in fear conditioning, are using unpleasant reinforcements like electric shock or other type of stressful influences. Thus, stress or traumatic experiences are those external cues that really can "talk to the genes." Another reason why transmission of memories was not noticed is that individual development and timing of exposure or conditioning was not taken into consideration. Most of Pavlov's animals were adult and mature. On the contrary, Pavlov's positive training may be also very important so far as it may lead to establishment of resilience, but in an appropriate time window.

We have deliberately avoided the topic of stress in this chapter to concentrate and outline most general aspects of modern understanding of epigenetics. Our further talk will be about stress, its severity, timing, nature and origin, and involvement of epigenetic mechanisms in programming stress responsivity of the organism. Our aim is to show how life stress, especially not only in the early periods of development, but also later in life, can shape vulnerability, resilience, mental health, disorders, and behaviors by utilizing epigenetic mechanisms. We will also discuss types of stressors and will deliberately focus on psychosocial stress that is especially relevant in modern societies.

References

Amasino, R. (2004). Vernalization, competence, and the epigenetic memory of winter. *The Plant Cell, 16*, 2553–2559.

Badcock, C., & Crespi, B. (2008). Battle of the sexes may set the brain. *Nature, 454*, 1054–1055.

Barber, R., Plumb, M. A., Boulton, E., Roux, I., & Dubrova, Y. E. (2002). Elevated mutations rates in the germ line of first and second generations offspring of irradiated male mice. *Proceeding of the National Academy of Sciences of the United States of America, 99*, 6877–6882.

Berger, S. L., Kouzarides, T., Schickhattar, R., & Shilatifard, A. (2009). An operational definition of epigenetics. *Genes and Development, 23*, 781–783.

Bernstein, B. E., Meissner, A., & Lander, E. S. (2007). The mammalian epigenome. *Cell, 128*, 669–681.

Burggren, W. (2016). Epigenetic inheritance and its role in evolutionary biology: re-evaluation and new perspectives. *Biology (Basel), 5*(2), pii: E24.

Chow, J., Yen, Z., Ziesche, S., & Brown, C. (2005). Silencing of the mammalian X chromosome. *Annual Review of Genomics and Human Genetics, 6*, 69–92.

Crews, D. (2008). Epigenetics and its implication for behavioral neuroendocrinology. *Frontiers in Neuroendocrinology, 29*, 344–357.

Damiani, G. (2007). The Yin and Yang of anti-Darwinian epigenetics and Darwinian genetics. *Rivista di Biologia, 100*, 361–402.

Dolinoy, D. C., Weidman, J. R., & Jirtle, R. L. (2007). Epigenetic gene regulation: linking early development environment to adult disease. *Reproductive Toxicology, 23*, 297–307.

Dupont, C., Armant, D. R., & Brenner, C. A. (2009). Epigenetics: definition, mechanisms and clinical perspective. *Seminars in Reproductive Medicine, 27*, 351–357.

Fraga, M. F., Ballestar, E., Paz, M. F., Ropero, S., Setien, M., Ballestar, M. L., Heine-Suñer, D., Cigudosa, J. C., Urioste, M., Benitez, J., Boix-Chornet, M., Sanchez-Aguilera, A., Ling, C., Carlsson, E., Poulsen, P., Vaag, A., Stephan, Z., Spector, T. D., Wu, Y. -Z., Plass, C., & Esteller, M. (2005). Epigenetic differences arise during the lifetime of monozygotic twins. *Proceeding of the National Academy of Sciences of the United States of America, 102*, 10604–10609.

Gluckman, P. D., & Hanson, M. A. (2004). Living in the past: evolution, development, and patterns of disease. *Science, 305*, 1733–1736.

Goldberg, A. D., Allis, D., & Bernstein, E. (2007). Epigenetics: a landscape takes shape. *Cell, 128*, 635–638.

Golubovskii, M. D. (2000). *Vek genetiki. Evolutsiya idey i ponyatiy (Century of genetics: Evolution of ideas and concepts)*. St. Petersburg: Borei Art.

Graham, L. R. (1987). *Science, philosophy, and human behavior in the Soviet Union*. New York: Columbia University Press.

Griffith, J. S., & Mahler, H. R. (1969). DNA ticketing theory of memory. *Nature, 223*, 580–582.

Groth, A., Rocha, W., Verreault, A., & Almouzni, G. (2007). Chromatin challenges during DNA replication and repair. *Cell, 128*, 721–733.

Hales, C. N., & Barker, D. J. P. (2001). The thrifty phenotype hypothesis. *British Medical Bulletin, 60*, 5–20.

Hochberg, Z. (2012). *Evo-devo of child growth: Treatise on child growth and human evolution*. New York: Wiley.

Hochberg, Z., Feil, R., Constancia, M., Fraga, M., Junien, C., Carel, J. -C., Boileau, P., Le Bouc, Y., Deal, C. L., Lillycrop, K., Scharfmann, R., Sheppard, A., Skinner, M., Szyf, M., Waterland, R. A., Waxmann, D. J., Whitelaw, E., Ong, K., & Albertson-Wikland, K. (2011). Child health, developmental plasticity, and epigenetic programming. *Endocrine Reviews, 32*, 159–224.

Holden, C. (2008). Parcing the genetics of behavior. *Science, 322*, 892–895.

Holliday, R. (1987). The inheritance of epigenetic defects. *Science, 238*, 163–170.

Holliday, R. (1994). Epigenetics—an overview. *Developmental Genetics, 15*, 453–457.

Holliday, R. (2006). Epigenetics. A historical overview. *Epigenetics, 1*, 76–80.

Hotchkiss, R. D. (1948). The quantitative separation of purines, pyrimidines and nucleosides by paper chromatography. *Journal of Biological Chemistry, 175*, 315–332.

Jablonka, E., & Raz, G. (2009). Transgenerational epigenetic inheritance: prevalence, mechanisms, and implications for the study of heredity and evolution. *The Quarterly Review of Biology, 84*, 131–176.

Jaenisch, R., & Bird, A. (2003). Epigenetic regulation of gene expression: how the genome integrate intrinsic and environmental signals. *Nature Genetics, 33*, 245–254.

Junko, A. A., Shaomin, Li., Hartley, D. M., & Feig, L. A. (2009). Transgenerational rescue of a genetic defect in long-term potentiation and memory formation by juvenile enrichment. *Journal of Neuroscience, 29,* 1496–1502.

Kanherkar, R. R., Bhatia-Dey, N., & Csoka, A. B. (2014). Epigenetics across the life span. *Cell and Developmental Biology, 2,* 49.

Le Bouc, Y., Rossignol, S., Azzi, S., Steunou, V., Netchine, I., & Gicquel, C. (2010). Epigenetics, genomic imprinting and assisted reproductive technology. *Annales d'Endocrinologie, 71,* 237–238.

Liu, Y., Liu, W. B., Liu, K. J., Ao, L., Zhong, J. L., Cao, J., & Liu, J. Y. (2015). Effect of 50 Hz extremely low-frequency electromagnetic fields on the DNA methylation and DNA methyltransferases in mouse spermatocyte-derived cell line GC-2. *BioMed Research International, 2015,* 10, 237183.

Lollicrop, K. A., Phillips, E. S., Jackson, A. A., Hanson, M. A., & Burdge, G. C. (2005). Dietary protein restriction of pregnant rats induces and folic acid supplementation prevents epigenetic modification of hepatic gene expression in the offspring. *Journal of Nutrition, 135,* 1382–1386.

Ma, S., Liu, X., Jiao, B., Yang, Y., & Liu, X. (2010). Low-dose radiation-induced responses: focusing on epigenetic regulation. *International Journal of Radiation Biology, 86,* 517–528.

Mandizabal, I., Keller, T. E., Zeng, J., & Soojin, V. Yi. (2014). Epigenetics and evolution. *Integrative and Comparative Biology, 54,* 31–42.

Marczylo, E. L., Amoako, A. A., Konje, J. C., Gant, T. W., & Marczylo, T. H. (2012). Smoking induces differential miRNA expression in human spermatozoa: a potential transgenerational epigenetic concern? *Epigenetics, 7,* 432–439.

Meaney, M., & Szyf, M. (2010). The seductive allure of behavioral epigenetics. *Science, 329,* 24–27.

Messerschmidt, D. M., Knowles, B. B., & Solter, D. (2014). DNA methylation dynamics during epigenetic reprogramming in the germline and preimplantation embryos. *Genes and Development, 28,* 812–828.

Milot, E., Mayer, F. M., Nussey, D. H., Boisvert, M., Pelletier, F., & Réale, D. (2011). Evidence for evolution in response to natural selection in a contemporary human population. *PNAS, 108,* 17040–17045.

Nazarov, V. I. (2007). *Evolutsiya ne po Darwinu. Smena evolutsionnoy modeli (Evolution not according to Darwin. Change of the evolutionary model)* (2nd ed.). Moscow: LKI.

Pawlow, I. P. (1923). New researches on conditioned reflexes. *Science, 58,* 359–361.

Pennisi, E. (2001). Behind the scenes of gene expression. *Science, 293,* 1064–1067.

Ramachandran, S., & Henikoff, S. (2015). Replicating nucleosomes. *Science Advances, 1,* e1500587.

Reik, W., & Walter, J. (2001). Genomic imprinting: parental influence on the genome. *Nature Reviews Genetics, 2,* 21–32.

Richards, E. J. (2006). Inherited epigenetic variation—revisiting soft inheritance. *Nature Reviews Genetics, 7,* 395–400.

Rozanov, V. A. (2013a). Epigenetics: stress and behavior. *Neurophysiology, 44,* 332–350.

Rozanov, V. A. (2013b). *Human ecology (chosen chapters).* Odessa: VMV.

Skinner, M. K., Amway, M. D., Savenkova, M. I., Gore, A. C., & Crews, D. (2008). Trasgenerational epigenetic programming of the embryonic testis transcriptome. *Genomics, 91,* 30–40.

Szyf, M. (2009a). The early life environment and the epigenome. *Biochimica et Biophysica Acta, 1790,* 878–885.

Szyf, M. (2009b). Epigenetic control of gene expression. The early life environment and the epigenome. *Biochimica Biophysica Acta, 1790,* 878–885.

Szyf, M. (2011). DNA methylation, the early-life social environment and behavioral disorders. *Journal of Neurodevelopment Disorders, 3,* 238–249.

Szyf, M. (2013). How do environment talk to genes? *Nature Neuroscience, 16*, 2–4.

Tammen, S. A., Friso, S., & Choi, S. -W. (2013). Epigenetics: the link between nature and nurture. *Molecular Aspects of Medicine, 34*, 753–764.

Tokunaga, M., Takahashi, T., De Meester, F., & Wilson, D. W. (2013). Nutrition and epigenetics. *Medical Epigenetics, 1*, 70–77.

Vaiserman, A. M. (2013). Long-term health consequences of early life exposure to substance abuse: an epigenetic perspective. *Journal of Developmental Origins of Health and Disease, 4*, 269–279.

Vaiserman, A. M. (2015a). Epigenetic programming by early-life stress: evidence from human populations. *Developmental Dynamics, 244*, 246–265.

Vaiserman, A. M. (2015b). Epidemiologic evidence for association between adverse environmental exposures in early life and epigenetic variation: a potential link to disease susceptibility? *Clinical Epigenetics, 7*, 96.

Vaiserman, A. M., Voitenko, V. P., & Mekhova, L. V. (2011). Epigeneticheskaya epidemiologiya vozrast-zavisimyh zabolevaniy (Epigenetic epidemiology of age-dependent diseases). *Ontogenes (Ontogenesis), 42*, 1–21.

Vanyushin, B. F. (2006). DNA methylation and epigenetics. *Russian Journal of Genetics, 12*, 985–997.

Vanyushin, B. F., Tkacheva, S. G., & Belozersky, A. N. (1970). Rare bases in animal DNA. *Nature, 225*, 948–949.

Wolff, G. L., Kodell, R. L., Moore, S. R., & Cooney, C. A. (1998). Maternal epigenetics and methyl supplements affect *agouti* gene expression in A^{vy}/a mice. *FASEB Journal, 12*, 949–957.

Wu, C. T., & Morris, J. R. (2001). Genes, genetics and epigenetics: a correspondence. *Science, 293*, 1103–1105.

Zhang, L., Hou, D., Chen, X., Li, D., Zhu, L., Zhang, Y., Li, J., Bian, Z., Liang, Z., Cai, Z., Yin, Y., Wang, C., Zhang, T., Zhu, D., Zhang, D., Xu, J., Chen, Q., Ba, Y., Liu, J., Wang, Q., Chen, J., Wang, J., Wang, M., Zhang, Q., Zhang, J., Zen, K., & Zhang, C. -Y. (2011). Exogenous plant MIR168a specifically targets mammalian LDLRAP1: evidence of cross-kingdom regulation by microRNA. *Cell Research (2012), 22*, 107–126.

Zhivotovsky, L. A. (2014). *Neizvestnyi Lysenko (An unknown Lysenko)*. Moscow: KMK.

4

Biological Embedding—How Early Life Stress Shapes Behaviors Later in Life and How Vulnerability is Built

Epigenetic transformations are responsible for many important biological processes, but especially for programming behaviors and diseases in response to different types of stressful situations. Some of these situations, such as nutritional or ecological stress were already discussed in the previous chapter, but the variety of these situations is much wider and includes not only signals from chemical and physical, but also from social environment, which is especially interesting in the context of topic we are discussing. Understanding of social environment may be very wide—it may include microsocial interaction with mother or both parents at the very early stages of life during realization of parental behavior, and later—with peers, rivals, enemies and mates, as well as with social structures in general, with their features, such as work load, unemployment, insecurity, inequalities, social exclusion, etc. For many animals early social and microsocial interaction are of utmost importance for development and future functioning. In humans some of these interactions may have a lot in common with other mammals at the most early stages of life, but may become much more complicated as the child grows, when personality and more complex psychosocial factors start to act. This development is often conceptualized as a life-course or life-trajectory model, which tries to integrate biological, psychological, and social factors and includes all stages of the individual life cycle, starting from preconception period to older age (Beddington et al., 2008; Alwin, 2012).

In agreement with life-trajectory model, an important concept of biological embedding is developed, which considers how life experiences, especially of early life periods (neonate, toddler, early childhood), and of

Stress and Epigenetics in Suicide. http://dx.doi.org/10.1016/B978-0-12-805199-3.00004-X

later life (adrenarche, puberty) gets "under the skin" through biological mechanisms associated with stress. It is important that it happens within specific time windows—during development and in critical or sensitive periods of life of the developing organism (Hertzman, 2012). This concept recently has received a strong support from studies of developmental origin of health and disease and behavioral epigenetics, making the whole picture more or less complete and logical. These ideas are very important from the point of view of our conceptual framework that aims to build an integrative picture of the role stress and epigenetics in suicide as of a tragic event that terminates life at certain point.

In mammals many physical (warmth, protection) and chemical (food, nutrition) environmental cues are tightly connected with microsocial interactions. For instance, in the early postnatal life the importance of nutrition is very high, but so far as it is provided by mother, it is normally combined with maternal care. This care provides to the newborn several very important feelings—being not alone, not abandoned, and that the entire environment is safe and protected. On the contrary, if maternal care, which combines nutrition, warmth, and feeling of protection, is terminated for a sufficiently long period of time, it means fears and even horrors of being abandoned, left alone, starved, and possibly frozen to death or killed by enemies and other hostile forces. These profound fears—fear of hunger, starvation, fear of loneliness (loss of social support), and fear to die or to be lost in a hostile environment are severe stressors of the early life periods which are inherent to all mammals. These fears reflect reactions to evolutionary significant negative circumstances. Stress produced by such profound fears, in case they reach high severity or happen rather often, as it became known recently, cause detectable changes in epigenetic landscape touching many systems of the organism, but mostly the system of stress-reaction itself—the hypothalamic-pituitary-adrenal (HPA). Such changes are programming behavioral patterns, traits, and physiological reactions that prepare an organism to survive in a harsh, hostile, and dangerous environment and include adaptive features, such as aggression, impulsivity, constant vigilance, and enhanced stress-reactivity. On the contrary, if maternal care is efficient and protective, the organism may be programmed to live in stable and friendly social environment. In accordance with the concept of ontogenetic programming which was discussed in Chapter 3, if environmental conditions in reality correspond programmed traits and behavioral patterns, it may ensure adaptive perspectives—better survival and more chances for reproduction. On the other hand, if environment does not match expectations, enhanced neuroendocrine system functioning and specific behavioral patterns programmed by early stress may result in behavioral and emotional abnormalities further in life. Such abnormalities in humans belong to the domain of mental ill health, and this field of study has recently been defined as behavioral and psychiatric epigenetics. Many animal models have been developed recently and many studies have helped to identify epigenetic events that underlie such programming. They suggest explanations for many epidemiological observations and promotes search for similar epigenetic findings in humans, which in many cases are confirming animal results, thus building a logical and persuasive picture of sociopsychobiological correlations.

EARLY LIFE STRESS AND BEHAVIORAL EPIGENETICS—ANIMAL MODELS

Today a great number of studies in rodents and other animals have brought significant body of evidence that perinatal stress including maternal distress, separation with mother, and variations in maternal care leads to stable neuroendocrine system perturbations in pups, which in turn, later in life may result in abnormal stress-reactivity, cognitive and emotional impairment, and other disturbances. Stress in utero is one of the most serious situations that may have long-lasting consequences, most of the earlier mentioned negative effects in later life are caused by high cortisol level that reaches neuroendocrine organs and particularly brain of the fetus or newborn and transmit stress signals into epigenetic marks (Hunter, 2012; Reynolds, Labad, Buss, Ghaemmaghami, & Raikkonen, 2013). In humans, high level of stress hormones in mothers' organism may have considerable effect on the fetus even despite the existence of placental "enzymatic barrier"—11β-hydroxysteroid dehydrogenases family. It is estimated that from 10% to 20% of maternal cortisol may reach the fetus in spite of this enzymatic activity which converts active cortisol into more inert cortisone (Meaney, Szyf, & Seckl, 2007). Excessive corticosteroids exposure affects brain development. It is well established, that in rodents, as well as in primates and humans, glucocorticoids are involved in the basic neuronal processes and brain development. During the prenatal period, rapid neurogenesis, cells migration and neuronal differentiation, dendritic arborization, axonal elongation, synapse formation, collateralization, synaptic pruning, and myelination take place. Increased levels of stress hormones may cause abnormal maturation of the brain myelination, structuring of synaptic connections, neuron-glia interrelations, and formation of the brain vasculature (McEwen, 1999).

Of course most of the results come from animal studies. For example, young and adult rats subjected to stress in utero demonstrate decreased densities of mineralocorticoid and glucocorticoid receptors in the hypothalamus, which may be important for the entire regulation of the HPA activity (Lupien, McEwen, Gunnar, & Heim, 2009). It may mean either abnormally intensified activity of this neuroendocrine system or its impaired regulation due to compromised feedback mechanisms

(Welberg, Seckl, & Holmes, 2001; Weinstock, 2005). Early studies have identified that these effects are associated with inhibition of the expression of glucocorticoid receptor (GR) in the hippocampal structures (Levitt, 1996). Later in life in utero stress-exposed animals with such peculiarities of regulation of the HPA exhibit a number of physiological and behavioral abnormalities, such as higher blood pressure, abnormal activity, impaired reaction to novelty and stress. Many of these symptoms resemble stress-related diseases and mental health problems in humans, including anxiety and depression (Weinstock, 2008; McEwen, 2008; Nestler, Pena, Kundakovic, Mitchell, & Akbarian, 2015).

Structural, metabolic, and functional disorders caused by stress hormones in developing organism are not limited by impaired regulation of HPA. Stress in utero also leads to changes in the dynamics of such important neuro-hormone as oxytocin, which may result in impaired social interactions, including parenting behavior (Lee, Brady, Shapiro, Dorsa, & Koenig, 2007). Prenatal stress also affects expression of NMDA glutamate receptors in the hippocampus, amygdala, and prefrontal cortex (Kinnunen, Koenig, & Bilbe, 2003) and activates expression of the corticotropin-releasing factor (CRF) gene in the amygdala (Welberg et al., 2001). The latter may be crucial, because amygdala CRF is an important factor of activation of the HPA and stress-reactivity associated with anger, aggression, and fear depicting involvement of amygdala in development of emotional reactions related to social domination or submission. Amygdala is often hyperactivated and even structurally more developed in animals bred up in stressful environment. Finally, impaired neuronal interactions in the limbic system and cortical structures of animals subjected to stress in utero can lead to increased sensitivity to alcohol and narcotic drugs. Some negative consequences, such as impairment of learning abilities and memory formation become more evident in the course of aging (Lupien et al., 2009). Results of other studies showed that adult rats subjected to stress in utero are characterized by disorders in the system of neuroendocrine control of the gonads and by deviations in sexual behavior. In particular, such males can demonstrate homosexuality, while in females lower fertility is observed. These disturbances are related to stress-induced changes in neurons of the anterior and mediobasal hypothalamic nuclei involved in neuroendocrine regulation of sexual behavior and sexual differentiation (Reznikov, Nosenko, Tarasenko, Sinitsyn, & Polyakova, 2001).

This short listing of brain structural and functional disturbances caused by stress in utero gives an impression how important this mechanism may be for programming behavioral abnormalities and corresponding mental health problems later in life. There are many evidences that epigenetic mechanisms, in particular DNA methylation and histones modifications, are the key mechanisms that establish long-lasting transcription profile of some critical genes belonging to HPA in stress (Meaney et al., 2007;

Champagne, 2013). Main targets for these modifications are receptors of different HPA regulators, mainly glucocorticoids, while results may vary in relation to period of gestation and sex of the offspring. For instance, very early prenatal exposure to mild nonpainful stress resulted in male (but not female) offspring in maladaptive behavioral stress-responsivity, behavioral abnormalities, and anhedonia. Higher stress-sensitivity in these mice was associated with long-term alterations in central CRF and GR expression, as well as increased HPA axis responsivity. These altered genes expression correlated with CRF and GR genes methylation, providing important evidence of epigenetic programming by prenatal stress (Mueller & Bale, 2008). What is mostly important, in cases of stress in utero abnormal epigenetic-dependent phenotypes can be registered in subsequent generations. For instance, second-generation offspring of prenatally stressed males rats had pecific gene expression patterns in nerve tissue during the perinatal sensitive period, when gonadal hormones due to their organizational effects establish the sexually dimorphic brain. Their male offspring showed a female-like pattern of genes important in neurodevelopment while miRNAs profile in their brain was changed. This was associated with morphological abnormalities of anogenital region in these male rats (Morgan & Bale, 2011). This study provides important example how effects of prenatal stress may be traced in male lineage in several generations. These results are also interesting in view of sex differences of some neurodevelopment disorders, such as autism and with regards of possible role in sexual dimorphism or brain and behavior.

It is important to mention that in many recent experimental studies, pregnant rats were subjected to different types of mild stressors, such as constant moderate noise, alternating light, and predator's odors, which make stress models much closer to natural situations (constant moderate stress with signs of psychosocial stress). In general the models of stress used in animal studies today are diverse and can be classified as acute stress, which includes restrain, forced swim and novelty exposure, and chronic stress, which includes chronic restrain, chronic social stress, mental, and pain stress. All these models are associated with epigenetic events in the central nervous system (Stankiewicz, Swiergiel, & Lisowski, 2013). It is also important to notice that most marked stress programming in utero can be seen within specific time window, early period of gestation being the most vulnerable one. Outside these sensitive (or critical) periods stress may not produce specific epigenetic marks and functional results (Stankiewicz et al., 2013).

After birth, an animal is surrounded by much more diverse environments and influences. Early postnatal period, when maternal care becomes critical, gives another great example how epigenetic mechanisms are involved in behavioral programming. All abnormalities in this period of life can be referred as microsocial stress caused by maternal deprivation or variations of maternal care and nurturing. Early studies in this field stemmed from experiments with rodents handling. It was noticed that after experimenter took away pups for some procedures and then returned them back to the cage, dams started to lick them more intensively. It soon became clear that this is a reaction for the periods of absence of the pups or their exposure to human hands and odors. It was found that short-term (3–15 min) separation of newborn pups from their mothers leads to enhanced growth and stress-resilience in the course of maturation. Such resilience was found to be related to enhanced expression of GR in the hippocampus and prefrontal cortex and also of subunits of GABAA receptors and benzodiazepine receptors in noradrenergic structures of the nucleus tractus solitarius, locus coeruleus, and amygdala, which are important regulators of the brain level of CRF (Meaney, 2001). Thus, short stress (separation) with subsequent intense tactile contacts with mother provides a programming positive effect on the state of the neuroendocrine system of stress-reactivity and sensitivity. It was logical therefore to test if longer separation may have a negative effect.

Experimentation with prolonged separation from the mother and more severe stress (which may be referred as modeling of early traumatic experience associated with fears and horrors of being abandoned, or in some sense—of impaired child-maternal bonding) gave important results. Animals subjected to such influences in a comparatively short time window after birth demonstrated in adult age significantly higher reactivity of HPA system to the standard stress (higher cortisol output, impaired feed-back regulation), which was associated with functional changes in the corresponding neuroendocrine system components. In particular, receptor binding of cortisol in the hippocampus, hypothalamus, and frontal cortex was found to be noticeably lower, the concentration of CRF-encoding mRNA was greater, and the concentration of CRF itself was higher (Plotsky & Meaney, 1993; Liu, Caldji, Sharma, Plotsky, & Meaney, 2000). Thus, moderate early stress promotes resilience to stress, while severe negative influences are programming stress-vulnerability and enhanced stress-sensitivity. These landmark studies initiated investigation of the programming role of early life adversities, such as low maternal warmth or neglect, but also in a more severe form, such as parental loss or physical and emotional abuse. Such experimental approaches (in many aspects similar to an early trauma in children according to the concepts of most psychologists) were used by a number of authors. By now dozens of original and review papers have been published either providing confirmations to above presented data and discussing how early life stress shapes pathological emotional reactions and programs dysfunctional patterns of behavior (Roth, 2013; Rozanov, 2013; Hing, Gardner, & Potash, 2014; Vaiserman, 2015).

Another series of studies that concentrated on natural individual variations of the intensity of mothering behavior in rats also provide excellent example of early programming of stress-reactivity. Mothering in rodents is a deeply instinctive behavior, which includes periods of intensive licking of pups by mothers. Such typical nursing interaction of dams with litter occurs predominantly at the time when the dam arches her back, so this pattern of behavior was labeled as licking-grooming arch-back nursing. It appeared that variation in this type of parental behavior has a normal distribution in case of a sufficient number of animals in the examined samplings. This fact testifies of genetic background of this complex behavioral pattern. With regards of this variability, all rodent dams can be divided into subgroups with intense, midintense, and weak licking-grooming behavior (Devlin, Brain, Austin, & Oberlander, 2010). A number of studies showed that adult offspring of rats with intense grooming, in case they were subjected to intensive parental behavior within the first postbirth days, are characterized by mild reactions to stress, a greater number of GR in the hippocampus, and weaker expression of the CRF gene in the hypothalamic cerebral zone (Liu et al., 1997, 2000; Weaver et al., 2004). Reaction to administered stress-test measured by blood cortisol level in such animals is characterized by low basal level with steep rise and rather quick reversal to base line, which is a normal stress-response associated with moderate energy costs, adaptation, and resilience. Young animals from low licking-grooming mothers, that is, developing under conditions of less maternal warmth and support, demonstrated opposite peculiarities. Their cortisol base level was usually higher, while response to administered stress was blunted. The relocation of newborn rats from low licking-grooming mothers to mothers with a bright maternal instinct (cross-fostering) resulted in reversion of the phenotype, that is, the resilience of the HPA in such young rats was recovered. This is a clear indication of the epigenetic nature of this effect (Meaney et al., 2007).

Later epigenetic mechanism was confirmed on the molecular level. In a series of experiments, it was shown that in the rat hippocampi the GR gene had different profile of DNA methylation in its promoter part, and these differences were associated with intensity of mothering behavior. In rats nurtured by mothers with low licking-grooming behavior promoter sequences were more often methylated, while in their better nurtured counterparts methylation level was lower. This fact was in full agreement with differences in GR expression and differential physiological reactivity of the whole stress-system. Dissimilarities in the level of acetylation of histones in the chromatin composition were also found. And the most important—these differences appeared within early stages of postnatal development in a rather narrow time window (days) when newborn animals were fully dependent on their mothers and were preserved during the entire life period of the animals.

So, experiences of early life social interactions provided by maternal behavior modify the profile of methylation of the genes, closely related to realization of the biological mechanism of reactions to stress (Weaver et al., 2004; Meaney et al., 2007). All these findings were confirmed by observations of natural development of animals. Behaviorally offspring of high licking-grooming mothers appeared to be less fearful and showed lower CRF expression in amygdala and lowered noradrenaline responses to stress. Conversely, animals from mothers exhibiting low maternal care, showed enhanced fear reactions and abnormalities in memory formation and behavior that resemble cognitive deterioration in older age. This is in a very good agreement with the mechanisms of neuroendocrine regulation in stress and with accumulating neurotoxic effects of frequent excessive cortisol peaks or stable high cortisol level—typical neuroendocrine correlates of high stress-reactivity and impaired cortisol control. Such neurotoxic influences are especially dangerous for hippocampus neurons rich for cortisol receptors and involved in HPA feedback regulation (Szyf, Weaver, Champagne, Diorio, & Meaney, 2005; Meaney et al., 2007). This mechanism demonstrates how epigenetic programming within the early stages of ontogenesis may actually form a "trajectory" of the entire life cycle, so far as stress-reactivity is a very strong and basic reaction, which may have consequences both for biological and behavioral abnormalities through life, while loss of hippocampus neurons may result in cognitive impairment in aging. It must be noted that exact molecular mechanisms with which mothering behavior establishes epigenetic marks are not clear. There are data that licking by mother in the first week of life may involve enhanced serotonin turnover and expression of the transcription factor NGFI-A in hippocampi of pups, which is known to increase GR gene expression in neurons (Meaney et al., 2007). Thus, neurotransmitter pathways involved in realization of parenting behavior may act as trigger mechanism.

Variations in mother behavior are known for other animals also, including primates, and there is evidence about matrilineal transmission of such behavior, so it looks as a universal biobehavioral regularity. This is especially true regarding extreme manifestations, such as clear neglect of the needs of the child and even child abuse on the one hand, and hyperprotection or excessive maternal anxiety on the other hand (Champagne, 2008). A number of observations on rhesus macaques and vervet monkeys demonstrated that diametrically differing behavioral patterns of mothering clearly belong to definite genetic lines of the females of these animals (Mastripieri, 1998; Bardi & Huffman, 2002). In human studies, it has also been showed that the level of affection of a woman for her mother is a strong predictor of affection of her daughter for her mother, that is, a definite behavioral pattern is observable through three generations (Benoit & Parker, 1994; Pederson, Gleason, Moran, & Sandi, 1998). Such facts were usually understood as the result of social learning, but data on involvement of biological mechanisms in transgenerational transmission of mothering behavior opened novel approach to this issue. There is evidence that epigenetics may be strongly involved in transgenerational transmission of this behavioral pattern. It is clear that transmission of the maternal behavior to female offspring in rodents is possible due to very early contacts with newborn animals in a rather narrow time window. It is impossible to adopt or copy this behavioral model at the moment so far as pups are equipped with the most primitive reactions and reflexes. Moreover, the licking-grooming behavior in female offspring will become obvious only when they will give birth to their litter, that is, after a rather prolonged period of physical and behavioral maturation. It means that imprinted behavior must be maintained for a long time in an implicit form, appearing only in an appropriate situation, when external or internal signals make this behavior necessary. It may have certain evolutionary and adaptive importance so far as existence of differential reactivity to stress within the population may mean important variability that may help to adapt to different types of environment.

In search of possible mechanism of earlier mentioned programming and its transmission to next generation all existing knowledge on neurobiology of maternal behavior was used (Champagne, 2008). Attention was focused on the oxytocin system which is a known mechanism of interaction of a lactating female with litter and, in general, of social interactions and formation of affections in mammals including humans (Donaldson & Young, 2008; Lee, Macbeth, Pagani, & Young, 2009). Main physiological mechanisms are oxytocin binding in the medial preoptic nucleus and the release of dopamine in the nucleus accumbens (main center of appetent pleasure) that is registered before the beginning and during the grooming. Characteristics of both mentioned processes were found to be contrastingly dissimilar in animals with high or low maternal behavior (Champagne, 2008). Moreover, all involved neural circuits and regulators appeared to be associated with effects of estrogens—it was found that promoter region of the gene encoding synthesis of oxytocin in the medial preoptic nucleus neurons interacts with a complex of the estrogen receptors (ER). Expression of these receptors in the earlier mentioned nucleus differs significantly in resenting and caring mothers, and what is mostly important, these differences quite expectedly were associated with differential ER promoter methylation in this brain nucleus (Champagne et al., 2006; Curley, Champagne, Bateson, & Keverne, 2008). This concludes the physiological mechanism and makes the whole scheme logical and consistent.

This study is a bright example how very important instinctive behavior like mother care in rats may be transmitted in generations by reproducing the intensity of maternal licking-grooming pattern in a matrilineal manner by involving epigenetic mechanisms. It is one of the important examples of Lamarckian mechanism of inheritance, which also underlines special role of transmission from mother to her female offspring. Though proven only for rodents it has a potential for primates and humans. It is interesting to notice that offspring of low licking-grooming mothers also had changed profile in methylation of GR gene and exhibited impaired reaction to stress (Pan, Fleming, Lawson, Jenkins, & McGowan, 2014). From this point it is quite understandable how early life stress, programmed enhanced stress-reactivity, and poor mothering can overlap, propagating negative behavioral characteristics in the populations (Lomanowska, Boivin, Hertzman, & Fleming, 2015). Moreover, stress-reactivity pattern may be transmitted both to females and males. Thus, poor mothering behavior, while being transmitted in a matrilineal manner, enhances stress-reactivity both in male and female offspring, thus, in next generations neglecting and stress-vulnerable female rats are more likely to mate with stress-vulnerable male rats, making the chances of behavioral problems even higher.

Further studies have shown that epigenetic modifications in newborn animals exposed to early life stress are not limited with genes that encode components of the stress-reactivity system. Low maternal care or prolonged separation also promotes epigenetic changes in several other genes that are important for stress regulation, cognitive control, addiction, and maternal behavior. For instance, it was found that newborn rats subjected to low mother warmth and grooming develop enhanced methylation of the BDNF gene in their prefrontal cortex (Roth, 2013). This protein is known to be a key factor of interneuron contacts and synaptic dynamics regulation that plays important role as neurogenesis promoter and general neuroplasticity agent. It is abundant in hippocampus, neocortex and forebrain, regions mostly involved in dynamic neuroplasticity associated with memory and cognitive functions. Accordingly, BDNF is deeply involved in memory formation, brain structures development and integrative activity in brain due to interrelations with other neurotrophic factors affecting neuroplasticity in a wide manner. Other brain molecules that are touched by methylation within low licking-grooming model are opioid receptors, ER in female offspring and antiinflammatory cytokines in male offspring (Roth, 2013). Thus, main epigenetic modifications in brain after early stressful exposures, both in utero and shortly after births are concentrated around HPA and neuroplasticity factors (Hing et al., 2014). This is in a good agreement with mentioned earlier neurodevelopment disturbances associated with depression-like behaviors, impaired cognitive functioning and higher stress-reactivity and sensitivity in animals subjected to early stress. After this analysis of conclusive results from animal models it is interesting to review the role of early stress (starting with most early in utero influences, and further early postnatal, early childhood and preadolescence stress) in humans with regards of possible role of epigenetics in programming poor mental health in children.

EARLY LIFE STRESS AND EPIGENETIC PROGRAMMING OF BEHAVIOR AND MENTAL HEALTH IN HUMANS

All earlier mentioned situations with animals are suggesting similar role of early life adversities in humans with behavioral and emotional outcomes in the form of common mental health problems and disorders. Different types of early life stress—stress experienced by mother during pregnancy, or insufficient level of contacts of the newborn with his/her mother, child neglect and abuse are rather common situations in different human populations. Studies of nutritional stress—hunger and their long-lasting consequences and programming of propensity to main "diseases of civilization," such as diabetes, metabolic syndrome, obesity, and myocardial infarction have been described earlier. On the other hand humanity is experiencing such kind of stress less and less, while other types of stress, mostly associated with social factors, are becoming more prevalent. Frequent anxiety and depression, poor sleep, feeling of being trapped, feeling of insecurity, anxiety and fears regarding future, that are so common in modern societies, mean constant perceived stress that takes part in what is called a "reciprocal recursion"—situation when subjectively perceived social conditions start to interact with deep internal biological mechanisms (Cole, 2014). In many papers it was proposed that observed growth of common mental health problems in a modern society is associated with growing psychosocial stress and that epigenetic transformation may serve a biological link between stressful societies and mental health problems (McGowan & Szyf, 2010; Rozanov, 2013; Jawahar, Murgatroid, Harrison, & Baune, 2015; Mitchell, Schneper, & Notterman, 2015; Provencal & Binder, 2015). This kind of stress is overwhelming; it hits both sexes, women who are pregnant and those who are planning a child, those who are already nurturing their babies, as well as young children, teenagers, working population, and older people. Our aim is to propose an integrative model of the role of stress and epigenetic transformations in humans in such phenomenon as suicide with regards of biological, psychological, and social factors inherent to this type of stress. Here we start with early life stress experienced prenatally and in the postnatal period.

Recent studies provide compelling information that maternal stress, including anxiety and depression of mother, is associated with an increased risk of different behavioral and emotional problems in their children, which may emerge later in life (Monk, Spicer, & Champagne, 2012; Vaiserman, 2015). A number of studies show that if a pregnant woman has been experiencing stress or symptoms of depression and anxiety, her children are at elevated risk of different psychopathologies (Monk et al., 2012). This evidence comes from observation on human populations that have been trapped in natural disasters. For instance, a cohort of pregnant women and their children that appeared in a stressful situation after severe ice storm in Canada in 1998, which led to electricity failure and austere conditions in homes of about of 6 million of people was followed and evaluated. It was shown that exposure to stress in utero in this cohort resulted in impairment of cognitive and language development of children at age 2 and 8 years, which differed with respect of stress severity and trimester of pregnancy in which the disaster happened (King & Laplante, 2005). In another study authors have been grouping pregnant women on the basis of the number of severe negative life events they have experienced during pregnancy. It was found that daughters of women, who experienced high psychosocial stress during pregnancy, were performing worse on working memory task, especially when additionally administered hydrocortisone (Entringer et al., 2009). Studies on neurological and psychological consequences of prenatal stress are numerous but until recently it was not clear to what extend epigenetics is involved in development of later-in-life mental health problems in humans.

It is quite clear that search for epigenetic marks in human nervous tissue have limitations. The ultimate goal is identification of epigenetic modifications in the brain after stress, especially in hippocampus and prefrontal cortex, which will mean confirmation of the whole concept. Such studies are possible as autopsies, for instance, in case of suicides and other reasons of death. These studies will be discussed further (Chapter 5). On the other hand, genome-wide studies recently are discussing to what extend stresses has a potency to change epigenome in other tissues of the organism, so far as elevated cortisol is influencing all body structures. Many studies of newborn children were performed on genome-wide basis and with tissues that are easily available, such as cord blood white cells—lymphocytes and neutrophils, or other tissues, such as buccal epithelium.

One of the most well-known is the study by Oberlander et al. (2008) who reported increased methylation of GR promoter in leucocytes of cord blood of those newborns, whose mothers had symptoms of depression or anxiety during the third trimester of pregnancy. Increased GR gene methylation in leucocytes was also associated with increased salivary cortisol in these toddlers in response to standard stress at age of 3 months (Oberlander et al., 2008). This finding raised hopes regarding nonneural markers of stress-induced epigenetic transformations. In addition, another work of the same group have revealed decreased methylation of a promoter of the well-studied SLC6A4 gene encoding the serotonin transporter both in maternal leucocytes and leucocytes of newborns of mothers who suffered depressed mood in the third trimester (Devlin et al., 2010). These and several other studies of children born to women exposed to stress during pregnancy (home violence, anxiety, depression) confirmed that such situations can be translated into epigenetic changes and stress-related phenotypes in children and that these changes can be traced in different types of cells, not only in brain. Some of these studies have concentrated on the role of placenta genes and activity of 11β-hydroxysteroid dehydrogenase as the main barrier for mother cortisol. It seems that some critical genes in placenta may be also epigenetically programmed by mother's psychosocial stress thus providing additional barrier function, or on the contrary, exerting higher risk of cortisol exposure (Vaiserman, 2015). On the other hand, one of the most recent studies, which actually confirmed that maternal psychological distress can be traced in increased GR and BDNF genes methylation in buccal epithelium of infants, showed that maternal cortisol during pregnancy was not correlated with it. The lack of association between maternal cortisol and infant DNA methylation suggests that effects of maternal depression may not be mediated directly by glucocorticoids and some other biological mechanisms may exist that translate maternal stress to epigenetic marks in infants (Braithwaite, Kundakovic, Ramchandani, Murphy, & Champagne, 2015). This is a very interesting issue so far as exact mechanism how epigenetic marks are mediated by personal stress remains to a great extend obscure.

Human studies are also providing compelling data that poor or inadequate maternal style may have biological epigenetic consequences. Though such studies are based on questioning of people regarding their past negative life events, which may produce distorted results, it was possible to prove that some stable epigenetic modifications are associated with early mothering style. For instance, mother's reports on the parenting provided to their children and methylation profile of GR and macrophage migration inhibitory factor (involved in immune system responses) were found to be correlated in their children when they were from 20 to 28 years old (Bick et al., 2012). Similar results are provided by another study in which high methylation of GR gene has been detected in blood leucocytes of adults who were subjected to maltreatment or inadequate nurturing in childhood (Tyrka, Price, Marsit, Walters, & Carpenter, 2012).

Studies on early postnatal stress focus not only on bad parenting, but also on different types of severe negative life events in childhood. The role of early psychological trauma and negative life events in childhood (such as violence, abuse, death of a parent or divorce, psychiatric illness of a parent, parental neglect, etc.) as predictors of mental health problems later in life has been well established in clinical and epidemiological studies. Problems of mental health that are associated with early childhood adversities usually include substance abuse, depression, anxiety, antisocial disorder, psychosis, and self-aggression (Brent & Silverstein, 2013). The role of stress-reactivity system dysfunction in mental health problems in children who has been subjected to childhood adversities is also verified. For instance, in case of depression, which is one of the most prevalent mental health problems in modern youth, HPA abnormalities may manifest themselves in both elevated or lowered stable cortisol level. In both cases there are signs of dysregulation and impaired feedback mechanisms of HPA which are associated with epigenetic programming of GR gene (Lopez-Duran, Kovacs, & George, 2009). These dysregulations are persistent, found in grown-ups, and serve as a background for psychopathologies. For instance, in women with a history of childhood abuse increased HPA and autonomic responses were found even if they were not depressed. On the other hand, this effect was particularly robust in women with current symptoms of depression and anxiety and was multiplied in women with a history of childhood abuse and a current major depression diagnosis which implies more severe HPA receptors dysregulation (Heim et al., 2000). In support of this, in lymphocytes of young (11–24 years) persons evaluated for childhood adversities (parental physical, emotional and sexual abuse, witnessed physical violence toward parents or siblings, peer physical violence, physical neglect and emotional neglect) strong correlation between borderline and other psychological ill-health symptoms and GR gene methylation profile was found—a sign of lower transcription and a reason for enhanced level of operation of the whole system (Radtke et al., 2015).

Such changes can be observed not only in abused children but also in those deprived of maternal warmth, for instance, brought up in the orphanages and other institutions for parentless children. Though needs in nutrition, protection, and care are fulfilled, such children lack a very important component—having one unique person who can combine all these factors—mother. In a number of studies, it was shown that institutionalized children have hundreds of differentially methylated genes in their blood cells if compared with children brought up by biological parents. Most of the genes involved are associated with signaling pathways or immune system (Naumova et al., 2012). In another study it was shown that risk of development of stress-related psychiatric disorders is associated with demethylation of FKBP5 gene, an important regulator of the GR reactivity. This demethylation was linked to increased stress-dependent gene transcription followed by a long-term dysregulation of the HPA and immune system functioning (Klengel et al., 2013). FKBP5 gene has already been mentioned, it was previously identified as an important factor involved in gene-to-environment interaction in PTSD (Binder et al., 2008). This gene variation itself does not affect the risk, while combination of the gene variant and past child abuse were the key ingredients for the doubled PTSD symptoms if traumatic event occurred. Very logically the product of FKBP5 is a member of the immunophilin proteins family, which interacts functionally with corticoid receptor complexes and regulates cortisol binding to its receptor (Binder et al., 2008). There is also a series of important studies of McGowan and Labonte, in which epigenetic signs in GR and some other genes were found in autopsy brains of suicide victims which were abused in childhood (McGowan et al., 2009; Labonte

& Turecki, 2011). They will be discussed in more details in the next chapter when epigenetic findings in suicide will be the main topic.

Summarizing this part of our discussion it is necessary to say, that currently there is a plethora of studies that link together early life adversities and ill mental health. The key component of this linkage are epigenetic biological mechanisms that translate mother stress or stressful life events experienced by a child in early childhood into stable disturbances of neuroendocrine networks with development of high stress-reactivity of the individual. Most of the studies in this field are registering enhanced methylation profile of the GR gene, which leads to decreased expression and lower representation of this receptor in brain tissues, which in turn leads to disturbances in the whole system of HPA and its feedback mechanisms. This agrees very well with the observed enhanced general stress-reactivity, higher cortisol rates in response to stress, higher baseline cortisol level, etc. Such physiological features are often found in depression and PTSD, as well as in other mental disorders, while entire biological scheme suggest plausible explanation of these mental health problems (Schroeder, Krebs, Bleich, & Frieling, 2010; Lester, Conradt, & Marsit, 2013). On the other hand in some studies cortisol dynamics in main stress-related psychiatric disorders does not fit into this scheme. It is therefore very probable that HPA dysregulation is only one of the reasons between many. In confirmation there are many findings regarding epigenetic labeling of important genes, such as neurotrophins and genes, responsible for immune system reactions. Methylated promoters for HPA genes after stress are found in leucocytes and lymphocytes, which rise hopes for finding diagnostics to predict mental health problems that need time for development.

In general there is a clear understanding of the whole logical scheme in which epigenetic transformation caused by early life stress establishes enhanced stress-responsivity and stress-reactivity for the whole life (which actually means stress-vulnerability in other terms) and which serves the basis for future diseases, including behavioral disorders (Weaver et al., 2004; Meaney et al., 2007; Szyf et al., 2005; McGowan & Szyf, 2010; Rozanov, 2013; Vaiserman, 2015). Within this logical scheme it is important to take into consideration different interactions that occur in the long run, such as genes–environment interaction (G×E) and with regards of the importance of the timing of the effect, genes environment–time (G×E×T) interaction. Beginning with the landmark study of Caspi, Sugden, and Moffitt (2003) in which the role of serotonin transporter gene variant has been linked to depression and suicidality and the role of early life stress was established, several other studies have found the same types of interactions. Between them is already mentioned FKB5 gene which appeared important for PTSD development in subjects that were abused in childhood (Binder et al., 2008). These genes, which are involved in serotonin dynamics and HPA regulation, and several other genes encoding neurotrophins

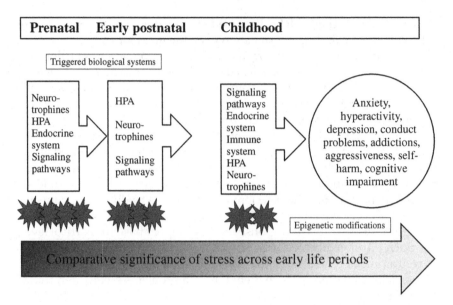

FIGURE 4.1   Neurodevelopmental pathways of mental health problems—role of early life periods.

that are important for the whole brain development, have been traced in several neuropsychiatric disorders, such as depression, PTSD, anxiety and addictions, which implies their nonspecific role in compromising brain structures in stress (Sapolsky, 2015). We will address these interactions in more details in the next chapter during discussion on integrative models of suicide. Meanwhile, we can make an interim conclusion, that early life stress in humans in the form of different adversities contributes greatly to the probability of neuropsychiatric disorders and mental health problems in the life course, and that epigenetic transformations induced by these adversities are possibly the main underlying mechanism that induce stress-vulnerability and propensity to ill mental health. Proposed pathogenetic pathways leading to behavioral and mental health problems accumulation are summarized in Fig. 4.1.

It is therefore very intriguing to what extend the role of epigenetics may be traced not only in cases of severe abuse and neglect of the child, but in a psychosocial stress throughout the whole life. The role of social factors (social interactions, socioeconomic factors, social defeats, and global failures) may be interesting to explore both in the early adulthood and later, across the life-span, so far as these factors are becoming more and more important when an individual reaches the status of the full member of the society with all problems and benefits provided by social environment and positive and negative influences that may appear on the life trajectory.

LIFE-TIME STRESS EXPERIENCES AND EPIGENETICS

We will start again with studies based on animal models so far as they provide most important and convincing information regarding epigenetic changes in brain, which is a critical organ within the whole scheme of HPA regulation and stress-induced neuroplasticity. Currently, these studies provide two main lines of evidence: based on the model of fear conditioning and on the model of social defeat. Both are very relevant for life-time stress and are close to natural social environments. Fear conditioning is a widely spread form of associative learning that may be studied in rodents, a variant of Pavlovian conditioned reflex. It is based on the combination of negative unconditioned (for instance, electrical foot shock) and conditioned signal (tone) or simply context (exposure to cage with electrical floor). The most relevant reaction of the animal is freezing, which is easy to register and measure (by duration). More extended outcomes are anhedonia and depression-like behavior with social withdrawal. Studies of fear conditioning have revealed several very interesting genes that undergo epigenetic modification during memorizing negative experiences (pain and fear of pain). For instance, in one study fearful memories consolidation is accompanied with demethylation and transcriptional activation of the reelin gene that encodes important memory enhancing glycoprotein, which regulates neuronal plasticity and long-term potentiation in hippocampus (Miller & Sweatt, 2007). It is interesting that such changes can be traced rather shortly after exposure to fearful stressor, within hours, showing how responsive and immediate DNA methylation process can be. It also returns us to the problem of possible triggers of epigenetic changes—could it be electrical events in the neurons that are linked together during learning and memorizing?

Except proteins and regulators directly associated with memory formation, DNA methylation changes within the BDNF gene in the hippocampus were revealed in this model (Lubin, Roth, & Sweatt, 2008). It is remarkable that these transformations were found in hippocampus which is known to be responsible for emotional memories. On the other hand, epigenetic traces were also found in the prefrontal cortex, where changed methylation of DNA associated with calcineurin—an important regulator of immune system—was identified (Miller et al., 2010). In other studies, it was also found that nonspecific histone deacetylase (HDAC) inhibition enhances fear-related learning and memory suggesting pivotal role of histone modifications in fear-related disorders, such as PTSD (Bowers, Xia, Carreiro, & Ressler, 2015). Direct facts that DNA methylation is involved in cognition and particularly memory formation and oblique evidence that histone acetylation enhancement improves learning and memory are of great interest. It may mean that memory can be boosted or suppressed by modulating molecular epigenetic mechanisms, which may have practical importance, and not only within the context of PTSD treatment.

It is also reasonable to mention in this context a well-known study by Lindqvist et al. (2007). In this study, chickens were subjected to unpredictable changes in lighting, which resulted in artificial modification of the circadian rhythm. Such situations are obviously stressful to the animal, but they resemble modern problems of millions of people way of leaving, with shift work, or frequent jet lags. In chicken, it resulted in the clear transgenerational effect, moreover, effect of domestication was registered. Offspring of stressed by lightning unpredictability domesticated chickens (but not wild ones) raised without parental contact (no possibility to transfer stress behaviorally) had a reduced spatial learning ability compared to offspring of nonstressed animals in a similar test as that was used for their parents. Offspring of stressed animals were also more competitive and grew faster than offspring of nonstressed parents. In a whole-genome cDNA microarray, the same changed hypothalamic gene expression profile caused by stress was found in the parents and in the offspring. This may mean that some memories (in this particular case, memories about day light cycle changes that spoil many adaptive behavioral patterns), can be transmitted in generations, again making the ghost of Lamarck appear at the stage. As the authors mention, the ability to transmit epigenetic information and behavior modifications between generations have been favored by domestication. In view of this it is interesting to remind about experiments of Dmitriy Belyaev from Novosibirsk who decades ago have discovered that domestication of wild fur animals, such as sable and mink leads to easier changes in their fur colors. It may mean that domestication may somehow enhance epigenetic transformations of different nature, being itself an example of fast evolutionary process that involves both classical behavioral genetics (selection of desired properties) and epigenetics with implications to behavior and stress-reactivity. In general changes in patterns of gene expression based on epigenetic programming are thought to be a leading mechanism of domestification (Jensen, 2015).

Social defeat stress is another animal model that makes studied of life time stress very close to human life situations. This model is quite well developed and also has a standardized protocol. The idea of this model is that male mouse is repeatedly subjected to several subsequent contacts with different larger and aggressive male mice. Animals that may serve as aggressors are selected in the preliminary experiments. An experimental animal is placed in the same cage with aggressive mouse, which leads to a series of fights in which an animal has no chance to win, after which the cage is divided with a septum that permits perception of the odor and appearance of the enemy. After series of defeats from several consequent aggressors the stressed animal develops depressive-like syndrome, characterized by anhedonia, lower locomotor activity, and deficits in social interactions (social avoidance). It is associated with neuroplasticity changes in the hippocampus and some electrophysiological signs that remain stable after the procedure is over (Buwalda et al., 2005). It was shown that this behavioral pattern is associated with a lasting decrease in methylation of CRH gene in the hypothalamus of animals (Elliott, Ezra-Nevo, Regev, Neufeld-Cohen, & Chen, 2010). On the other hand, some mice remain more or less intact even after many exposures to this type of stress. The most remarkable fact is that these mice did not show these epigenetic modifications also, thus allowing to discuss possible mechanisms of resilience to social defeat stress.

It is very remarkable that epigenetic modifications after stress are found in hippocampus—an organ of emotional memory, containing many receptors for glucocorticoids and neurons, which are susceptible to hypoxia. Alterations in hippocampal DNA methylation are found in different chronic and acute stress models (Roth, 2013). One of the models based on combination of acute and chronic stress was aimed to produce PTSD-like symptoms in animals. The acute component included two sessions of 1 h immobilization during predator (cat) exposure, while the chronic component consisted of exposure of rats to unstable housing conditions for a month. Such combination resembles modern life with not very severe, but inescapable chronicity of stressors, which is complemented by unpredictable stressful life events. It was found that "psychosocially" stressed rats had reduced growth rate, lower thymus, and increased adrenal gland weight (classical signs of stress according to Seyle), and also increased anxiety, exaggerated startle response, cognitive impairments, and greater cardiovascular and corticosterone level reactivity to an acute test stressor (Zoladz, Conrad, Fleshner, & Diamond, 2008). Later it was shown that these behavioral abnormalities coincide with alterations in hippocampal BDNF gene methylation and gene expression (Roth, Zoladz, Sweatt, & Diamond, 2011).

In another study, a model of daily chronic ultramild stress was used. This model is based on several alternating conditions like staying in a tilted or soiled cage, in a cage smaller than usual, with night illumination or with periodical short time food restrictions. In this study inbred stress-susceptible and stress-resilient male mice developed differential patterns of glial cell-derived neurotrophic factor (GDNF) DNA methylation within the nucleus accumbens (Uchida et al., 2011). Such studies link together very mild daily variable stress, which is very close to social situations in humans with epigenetic transformations in a part of the brain that is involved in cognitive processing of aversion, motivation, pleasure, reward and reinforcement learning, and stress-vulnerability and resilience.

Overall, existing models of mild day-to-day stress and memorizing of fearful experiences provide evidence that epigenetic transformations are quite detectable in these situations. Their localization is logically linked to brain parts that are responsible for emotional memories, stress and anxiety (hippocampus and amygdala), while genes involved are coding products that play a role in memorizing and in brain cellular circuit's neuroplasticity. Interactions of amygdala and hippocampus with prefrontal cortex and recently confirmed as a mechanism of extinction or inhibition of conditioned fear (Kim & Jung, 2006). All this makes the whole picture rather logical and compelling.

Most of the studies cited above are utilizing a candidate gene approach, while recently results of genome-wide assays focused on nonneural tissues provide evidence for much wider set of genetic loci that are differentially methylated in stressful conditions. This approach suggests a lot of interesting ideas for further in depth investigation regarding molecular consequences of stress. It is most widely utilized in human studies in which the availability of the biological material is one of the limitation factors.

BIOLOGICAL EMBEDDING—HOW SOCIAL ENVIRONMENT "GETS UNDER THE SKIN"

Biological embedding is a comparatively new concept. It has been developed almost simultaneously with growing understanding of the role of epigenetics in health programming, especially mental health (Hertzman, 2012; Sasaki, de Vega, & McGowan, 2013). The basis for the concept are human epidemiological studies and animal studies that provide examples how early life negative and positive experiences influence health in a long-term manner. As it was already presented earlier, such experiences produce most profound outcomes if the damaging factor acts during periods of high brain plasticity in the prenatal period and during early postnatal life, though not limited by these periods. Hertzman specifies the content of biological embedding by focusing on its 4 main attributes: 1) experience "gets under the skin" by altering biodevelopmental processes; 2) systematic experiences from different social environments lead to systematically different states; 3) these differences remain stable for the whole life; and 4) these differences have the capacity to influence health, well-being, learning or behavior over the life course (Hertzman, 2012).

The biological embedding concept explains how social factors determine population health outcomes. Though health is the result of individual genetics, development, behavior, psychology and many other factors, social determinants of health play the most important role on the population level. As Hertzman argues, for the last 25 years, a priority factor in population health has been the socioeconomic gradient. This is true for health measured by mortality and morbidity, but even much more true for mental health (Friedli, 2009). The whole concept is in a very good agreement with the life trajectory concept and understanding of the role of epigenetic transformations induced by early life adversities. Lower socioeconomic status, poor social and economic circumstances obviously affect health, while role of early life conditions becomes more evident with progression through the life cycle. The list of possible risks and disadvantages includes having poor housing and bringing up children in less secure conditions, lower education of adults and children, poor diet and lower level of necessary nutrients, higher stress and depression rate in parents and consequent problems in children (Wilkinson & Marmot, 1998; Friedli, 2009). In many studies it was established that low socioeconomic status is associated with higher prevalence of psychoemotional disturbances, including depression, anxiety and psychosis (Hackman, Farah, & Meaney, 2010). Brain is the central neuroendocrine organ that plays pivotal role both in cognitive perception of stress and in realization of stress-reaction and at the same moment a target for toxic effects of stress hormones (McEwen & Gianaros, 2010). Brain actually suggests a decision what is stress and determines biological and behavioral responses that may be either deleterious or beneficial in terms of health and disease. Moreover, brain changes under stress itself and may direct activity of main systems of the organism (neuroendocrine, metabolic, cardiovascular, and immune) inducing either negative or positive outcomes. At certain moment of stress chronicity brain becomes main mediator between perceived internal stress in a form of social anxiety, depressive symptoms, aggressive and autoagressive impulses, sleep disturbances, pain, and outcomes of "wear and tear"—clinical depression, addictions, stroke, heart attack, or PTSD. Epigenetic transformations seem to be one of the important mechanisms of such development. It is an important moment for our modeling and a central point of the concept of psychosocial stress and its programming effects.

Quite logically several studies have been focused on possible epigenetic consequences of social adversities, and in particular, poverty. In one of such studies, an original approach was used. The analysis of the level of methylation of the entire genome was carried out in 45-year-old representatives of diametrically differing socioeconomic status (SES) groups. Individuals belonging to groups with different SES at present and those belonging to such groups from birth were compared. About 20,000 genes and 400 microRNA were included to the analysis. Differences in the level of methylation of promoters were found in more than 6000 genes. In those subjects, whose childhood was spent under differing socioeconomic conditions, much more differences in methylation sites (1252 genes) were found. Subjects, who belonged to similar classes in childhood, but who occupied contrasting positions in the social hierarchy within subsequent years demonstrated only 545 differences in promoter regions (Borghol et al., 2012). Functionally most genes that were found to be differentially methylated belonged to various metabolic and cell signaling pathways. These data emphasize that conditions imposed by early social position are more important for establishing marks on the genome than current social state.

In another more focused study, Miller and Chen (2007) have investigated mRNA for 2 candidate genes—GR and toll-like receptor 4 (TLR4) in peripheral blood leukocytes of 136 female adolescents in relation to such factor as house ownership. The study have reported that the participants, whose families owned homes during their childhood (years 2 to 3 of life appeared to be a critical period) showed higher GR mRNA and lower TLR4 mRNA during adolescence, a profile that suggests better regulation of inflammatory responses. These effects were not mediated by current economic circumstances, life stress, or health practices, moreover, changes in SES in later life were unable to "undo" these effects (Miller & Chen, 2007). Later the same authors have reported that subjects with low early-life socioeconomic status had significant up-regulation of genes, which are bearing response elements for the CREB/ATF family of transcription factors that conveys adrenergic signals to leukocytes, and significant down-regulation of genes with response elements for the glucocorticoid receptor, which regulates the secretion of cortisol and transduces its antiinflammatory actions in the immune system. Subjects from low-SES backgrounds also showed increased production and higher peaks of cortisol in daily life and greater stimulated production of the proinflammatory cytokine interleukin 6. These disparities were independent of subjects' current status, lifestyle practices, and perceived stress. Collectively, these data suggest that children reared in families with low SES obtain a defensive phenotype characterized by resistance to glucocorticoid signaling, which in turn facilitates exaggerated adrenocortical and inflammatory responses. Although these response patterns could serve as adaptive during acute stressful situations, in a longer perspective they might cause allostatic load on the body that, as the authors argue, may have long-lasting consequences ultimately contributing to the chronic diseases of older age, such as arthritis, systemic diseases, etc. (Miller et al., 2009).

In another study the same group of authors tried a hypothesis that adults who grew up in low SES households, but who experienced higher levels of maternal warmth would be protected from the above mentioned proinflammatory reactions. For this purpose 53 healthy adults (aged 25–40), all of them from low SES environments early in life, were assessed on markers of immune activation and systemic inflammation. Maternal warmth experienced in childhood was measured using the Parent Bonding Inventory. This inventory consists of a subscale assessing maternal warmth/care which includes 12 items depicting the quality of relationship with mother during childhood (e.g., "My mother spoke to me in a warm and friendly voice.," etc.). Genome-wide transcriptional analysis confirmed that low early life SES individuals who had high warmth expressing mothers, exhibited less production of interleukin 6, reduced proinflammatory transcription factor activity (NF-κB), and immune activating transcription factor activity (AP-1) compared to those who were

Risk factors   Epigenetics   Family functioning Outcomes

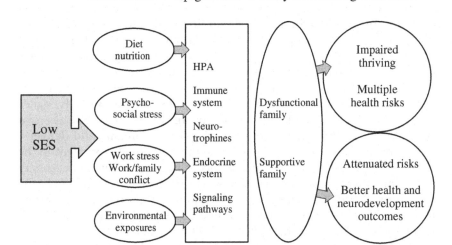

FIGURE 4.2 Epigenetic factors of socioeconomic status and associations with development and health outcomes.

low SES early in life, but experienced low maternal warmth (Chen, Miller, Kobor, & Cole, 2011). This is a remarkable study which gives an impression, how risks associated with low SES may be attenuated by parental involvement and stress-protective behavior. Recent review on the impact of poverty on brain development is discussing how neural plasticity, epigenetics, material deprivation (e.g., nutrient deficiencies), stress (e.g., negative parenting), and environmental toxins may shape the developing brain. Existing evidence for the relationship between child poverty and brain structure and function focus on brain areas that support memory, emotions regulation and higher-order cognitive functioning (hippocampus, amygdala, prefrontal cortex), as well as regions that support language and literacy (cortical areas of the left hemisphere) (Johnson, Riis, & Noble, 2016). Epigenetic transformations, which can establish certain cellular processes for the whole life and can potentially be transmitted to further generations are seen as central mechanism. Logical scheme that links together early life SES, existing risks, and level of parental support in relation to epigenetic events and health outcomes are presented in Fig. 4.2.

As can be seen, most studies that evaluate role of epigenetics in biological embedding focus on early life experiences so far as organism is more susceptible to trauma and supportive influences during sensitive periods that exist in childhood. At the same time, all further history of life of the individual (the level of experienced stress or social support across the life-span) can also participate in the formation of the "epigenetic landscape." Probably, such changes accompany the entire life trajectory, and their pattern can be rather important in certain life periods for increases in the stress-vulnerability or for building resilience of the personality. In this respect, social support, empathy, advise, or psychotherapy may be important protective events. In one of the interesting studies changes in DNA methylation immediately following a specific stressful event in humans were investigated (Unternaehrer et al., 2012). Study participants were adults aged 61–67 years who had been exposed to war adversities in early childhood. Standard stress exposure was the Trier Social Stress Test (TSST)—method that requires participants to prepare and deliver a speech and verbally respond to a challenging arithmetic problem in the presence of a socially evaluative audience. Unpredictability and necessity to present in front of an authoritative group are main factors that make the task stressful and at the same time very typical for social life. Each participant had blood samples drawn at a time point prior to the TSST, 10 min following completion of the TSST, and 90 min after the TSST. Examination of DNA methylation patterns of two stress-related genes—oxytocin receptor and BDNF, revealed DNA methylation changes across sequences of oxytocin receptor with less change detected in BDNF gene sequence. This study is of great importance to our mind so far as it shows that even a short stressful social event, and maybe even thoughts about social defeat or kind of humiliation associated with the inability to solve a task in the social circumstances may have immediate consequences in epigenetic profile of brain- and behavior-related genetic factors.

Associations between SES and epigenetic transformations provide an example how conservative biological mechanisms that are important for adaptation and possibly take part in biological evolution get involved in the effects of psychosocial stress. Recently, this topic has received strong support from the concept of "social genomics" (Cole, 2014). Human social genomics analyze how everyday life circumstances influence human genes expression. Recently, several studies have revealed that social–environmental conditions, such as urbanity, low socioeconomic status, social isolation, social threat, and low or unstable social status are associated with differential expression of hundreds of gene transcripts in leukocytes and diseased tissues, such as metastatic cancers. Analysis of the set of genes within the immune system gives an impression, that in socially isolated individuals upregulated genes include a set of transcripts that play a central role in inflammation, while downregulated transcripts include Type I interferon innate antiviral responses and antibodies against viruses. Such differential activation/inactivation of genes to a great extend explain links between modern urbanity and overcrowding and most prevalent diseases, like rise of inflammation (including neuroinflammation) and autoimmune disorders and low immune response to viral infections (Cole, 2013, 2014). This repetitive transcriptional repertoire is defined as a

Conserved Transcriptional Response to Adversity (CTRA) which is inherent to responses to physical threats historically associated with possible trauma and increased risk of infection. As to antiviral response suppression, its negative role possibly was not evident while people were living in small communities and became evident as a consequence of overcrowding in the cities. Moreover, besides immune system, which seems central to all social adversities, several studies have revealed relationships between social conditions and gene expression in diseased tissues, such as breast, ovarian, and prostate cancers. In these solid epithelial tumors low social support is associated with increased expression of many genes that promote cancer progression, metastasis, and resistance to anticancer drugs (Cole, 2014). As to neuroinflammation its association with psychosocial stressors is considered an important additional mechanism that leads to depression and memory impairment. Brain appeared very susceptible to immune attacks in spite of existing barriers, especially activation of microglia turned to be important factor that completes pathological development in psychosocial stress (McKim et al., 2016; Calcia et al., 2016) (Fig. 4.3).

The elegant concept of social genomics and CTRA is important for our discussion not only because it links together social conditions and typical human stresses, including low socioeconomic status, chronic stress (e.g.,

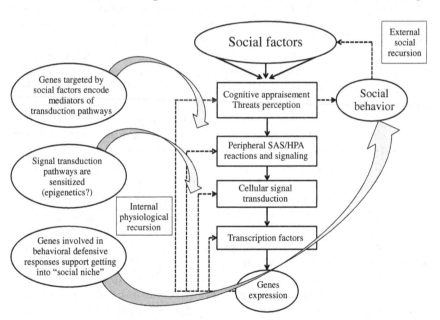

FIGURE 4.3 Social genomics pathways and role of reciprocal recursion. *Source: Adapted from Cole, S. W. (2014). Human social genomics. PLoS Genetics, 10(8), e1004601, by courtesy of the author.*

care-giving for a dying spouse), bereavement, posttraumatic stress disorder and cancer diagnosis with transcriptome, but also due to several interesting and important integrative ideas and explanations. The first one is related to the selection and presentation of facts that social, symbolic, or imagined threats occurring in the contemporary social environment can also activate CTRA, and due to chronicity of rumination and negative thinking it itself may become the factor of chronicity of disorders, such as arthritis, depression, neurodegenerative disorders, or cancer. Another very important component of the developing concept is that social-environmental threats may be cognitively appraised and converted into new pattern of activity of stress systems (HPA and SAS), which can lead to internal and external physiological recursion (Cole, 2014). Internal physiological recursion is understood as a mechanism when genes targeted by social signal transduction pathways start to propagate experience-induced transcriptional alterations by sensitizing the same transduction pathways to the external social environments. External social recursion is understood as mechanism by which social signals modulate genes involved in regulation in social behaviors. Combination of these mechanisms, as it is modeled by concept of social genomics, can lead to a programmed set of sickness behavior including reduced social motivation, fatigue, anhedonia, and depression (Cole, 2014). It also involves a reduction of individual social behavior and selection of altered social niche (group, company, community, behavioral pattern, etc.), which implies less social support and more hostile environment. This idea may be supplemented—in chronic stress with allostatic load an individual more easily gets involved in what is sometimes called allostatic behavior. It is typical behavior associated with chronic frustration and aimed on immediate rewards—drinking, sexual promiscuity, illicit drugs, or medicines that are soothing the symptoms of anxiety. This makes a dangerous vicious cycle in which epigenetic events also may be involved. On the other hand, the author of the concept of social genomics argues that internal and external recursion may happen even without requiring any epigenetic modifications, though leaves place for it in his pathogenetic scheme (Cole, 2013, 2014).

This type of thinking and such understanding of the role of epigenetics in modern psychosocial stress and possible role in propagating internal feeling of stress and unhealthy behaviors is very relevant for our integrative model of the role of stress and epigenetics in suicide. The question why in the modern society more and more people including adolescents and very young adults are experiencing more mental health problems and appear at higher risk of suicide can be resolved only on the basis of a complex biopsychosocial synthesis and in a modeling that encounters all possible interrelations between stress, epigenetics, neuroplasticity, cognitive processes, and social factors. This will be the subject of discussion in the next chapter.

References

Alwin, D. F. (2012). Integrating varieties of life course concepts. *The Journal of Gerontology, Series B, 67B*, 206–220.

Bardi, M., & Huffman, M. A. (2002). Effects of maternal style on infant behavior in Japanese macaques (*Macaca fuscata*). *Developmental Psychobiology, 41*, 364–372.

Beddington, J., Cooper, C. L., Field, J., Goswami, U., Huppert, F. A., Jenkins, R., Jones, H. S., Kirkwood, T. B., Sahakian, B. J., & Thomas, S. M. (2008). The mental wealth of nations. *Nature, 455*, 1057–1060.

Benoit, D., & Parker, K. C. (1994). Stability and transmission of attachment across three generations. *Child Development, 65*, 1444–1456.

Bick, J., Naumova, O., Hunter, S., Barbot, B., Lee, M., Luthar, S. S., Raefski, A., & Grogorenko, E. L. (2012). Childhood adversity and DNA methylation of genes involved in the hypothalamus-pituitary-adrenal axis and immune system: whole-genome and candidate-gene associations. *Developmental Psychopathology, 24*, 1417–1425.

Binder, B., Bradley, R. G., Wei, L., Epstein, M. P., Deveau, T. C., Mercer, K. B., Tang, Y., Gillespie, C. F., Heim, C. M., Nemeroff, C. B., Schwartz, C., Cubells, J. F., & Ressler, K. J. (2008). Association of FKBP5 polymorphisms and childhood abuse with risk of posttraumatic stress disorder symptoms in adults. *Journal of the American Medical Association, 299*, 1291–1305.

Borghol, N., Suderman, M., McArdle, W., Racine, A., Hallett, M., Pembrey, M., Hertzman, C., Power, C., & Szyf, M. (2012). Associations with early-life socio-economic position in adult DNA methylation. *International Journal of Epidemiology, 41*, 62–74.

Bowers, M. E., Xia, B., Carreiro, S., & Ressler, K. J. (2015). The Class I HDAC inhibitor RGFP963 enhances consolidation of cued fear extinction. *Learning and Memory, 22*, 225–231.

Braithwaite, E. C., Kundakovic, M., Ramchandani, P. G., Murphy, S. E., & Champagne, F. A. (2015). Maternal prenatal depressive symptoms predict infant NR3C1 1F and BDNF IV DNA methylation. *Epigenetics, 10*, 408–417.

Brent, D. A., & Silverstein, M. (2013). Shedding light on the long shadow of childhood adversity. *JAMA, 309*, 177–1778.

Buwalda, B., Kole, M. H. P., Veenema, A. H., Huininga, M., de Boer, S. F., Korte, S. M., & Koolhaas, J. M. (2005). Long-term effects of social stress on brain and behavior: a focus on hippocampal functioning. *Neuroscience & Biobehavioral Reviews, 29*, 83–97.

Calcia, M. A., Bonsail, D. R., Bloomfield, P. S., Selvaraj, S., Barichello, T., & Howes, O. D. (2016). Stress and neuroinflammation: a systemic review of the effects of stress on microglia and the implication for mental illness. *Psychopharmacology (Berl.), 233*, 1637–1650.

Caspi, A., Sugden, K., & Moffitt, T. E. (2003). Influence of life stress on depression: moderation by a polymorphism in the 5-HTT gene. *Science, 301*, 386–389.

Champagne, F. A. (2008). Epigenetic mechanisms and the transgenerational effects of maternal care. *Frontiers in Neuroendocrinology, 29*, 386–397.

Champagne, F. A. (2013). Early environment, glucocorticoid receptors, and behavioral epigenetics. *Behavioral Neurosciences, 127*, 628–636.

Champagne, F. A., Weaver, I. C., Diorio, J., Dymov, S., Szyf, M., & Meaney, M. J. (2006). Maternal care associated with methylation of the estrogen receptor-alpha1b promoter and estrogen receptor-alpha expression in the medial preoptic area of female offspring. *Endocrinology, 147*, 2909–2915.

Chen, E. E., Miller, G. E., Kobor, M. S., & Cole, S. W. (2011). Maternal warmth buffers the effects of low early-life socioeconomic status on pro-inflammatory signaling in adulthood. *Molecular Psychiatry, 16*, 729–737.

Cole, S. W. (2013). Social regulation of human gene expression: mechanisms and implications for public health. *American Journal of Public Health, 103*(Suppl. 1), S84–S92.

Cole, S. W. (2014). Human Social Genomics. *PLoS Genetics, 10*(8), e1004601.

Curley, J. P., Champagne, F. A., Bateson, P., & Keverne, E. B. (2008). Transgenerational effects of impaired maternal care on behavior of offspring and grand offspring. *Animal Behavior*, 75(4), 1551–1561.

Devlin, A. M., Brain, U., Austin, J., & Oberlander, T. F. (2010). Prenatal exposure to maternal depressed mood and the MTHFR C677T variant affect SLC6A4 methylation in infants at birth. *PLOS One*, 5(8), e12201.

Donaldson, Z. R., & Young, L. J. (2008). Oxytocin, vasopressin, and the neurogenetics of sociality. *Science, 322*, 900–904.

Elliott, E., Ezra-Nevo, G., Regev, L., Neufeld-Cohen, A., & Chen, A. (2010). Resilience to social stress coincides with functional DNA methylation of the crf gene in adult mice. *Nature Neuroscience, 13*, 1351–1353.

Entringer, S., Buss, C., Kumsta, R., Hellhammer, D. H., Wadhwa, P. D., & Wust, S. (2009). Prenatal psychosocial stress exposure is associated with subsequent working memory performance in young women. *Behavioral Neurosciences, 123*, 886–893.

Friedli, L. (2009). *Mental health, resilience and inequalities*. WHO Regional Office for Europe.

Hackman, D. A., Farah, M. J., & Meaney, M. J. (2010). Socioeconomic status and the brain: mechanistic insights from animal and human research. *Nature Reviews Neuroscience, 11*, 651–659.

Heim, C., Newport, D. J., Heit, S., Graham, Y. P., Bonsall, R., Miller, A. H., & Nemeroff, C. B. (2000). Pituitary-adrenal and autonomic responses to stress in women after sexual and physical abuse in childhood. *JAMA, 284*, 592–597.

Hertzman, C. (2012). Putting the concept of biological embedding in historical perspective. *PNAS, 109*, 1750–1757.

Hing, B., Gardner, C., & Potash, J. B. (2014). Effects of negative stressors on DNA methylation in the brain: implication for mood and anxiety disorders. *American Journal of Medical genetics. B. Neuropsychiatric Genetics, 165*, 541–554.

Hunter, R. G. (2012). Epigenetic effects of stress and corticosteroids in the brain. *Frontiers in Cellular Neuroscience, 6*, 18.

Jawahar, M. C., Murgatroid, C., Harrison, E. L., & Baune, B. T. (2015). Epigenetic alterations following early postnatal stress: a review of novel aetiological mechanisms of common psychiatric disorders. *Clinical Epigenetics, 7*, 122.

Jensen, P. (2015). Adding 'epi-' to behaviour genetics: implications for animal domestication. *Journal of Experimental Biology, 218*, 32–40.

Johnson, S. B., Riis, J. L., & Noble, K. (2016). State of the art review: poverty and developing brain. *Pediatrics, 137*(4), .

Kim, J. J., & Jung, M. W. (2006). Neural circuits and mehcanisms involved in Pavlovian fear conditioning: a critical review. *Neuroscience and Biobehavioral Reviews, 30*(2), 188–202.

King, S., & Laplante, D. P. (2005). The effects of prenatal maternal stress on children's cognitive development: project Ice Storm. *Stress, 8*, 35–45.

Kinnunen, A. K., Koenig, J. I., & Bilbe, G. (2003). Repeated variable prenatal stress alters pre- and postsynaptic gene expression in the rat frontal pole. *Journal of Neurochemistry, 86*, 736–748.

Klengel, T., Mehta, D., Anacker, C., Rex-Haffner, M., Pruessner, J. C., Pariante, C. M., Pace, T. W. W., Mercer, K. B., Mayberg, H. S., Bradley, B., Nemeroff, C. B., Holsboer, F., Heim, C. M., Ressler, K. J., Rein, T., & Binder, E. B. (2013). Allele-specific FKBP5 DNA demethylation mediates gene–childhood trauma interactions. *Nature Neuroscience, 16*, 33–41.

Labonte, B., & Turecki, G. (2011). The epigenetics of suicide: explaining the biological effect of early life environmental adversity. *Archives of Suicide Research, 14*, 291–310.

Lee, P. R., Brady, D. L., Shapiro, R. A., Dorsa, D. M., & Koenig, J. I. (2007). Prenatal stress generates deficits in rat social behavior: reversal by oxytocin. *Brain Research, 1156*, 152–167.

Lee, H. J., Macbeth, A. H., Pagani, J. H., & Young, W. S. (2009). Oxytocin: the great facilitator of life. *Progress in Neurobiology, 88*, 127–151.

Lester, B. M., Conradt, E., & Marsit, C. J. (2013). Epigenetic basis for the development of depression in children. *Clinical Obstetrics and Gynecology, 56,* 556–565.

Levitt, N. S. (1996). Dexamethasone in the last week of pregnancy attenuates hippocampal glucocorticoid receptor gene expression and elevates blood pressure in the adult offspring of the rat. *Neuroendocrinology, 64,* 412–418.

Lindqvist, C., Janczak, A. M., Natt, D., Baranowska, I., Lindqvist, N., Wichman, A., Lundeberg, J., Lindberg, J., Torjesen, P. A., & Jensen, P. (2007). Transmission of stress-induced learning impairment and associated brain gene expression from parents to offspring in chickens. *PLOS One, 2*(4), e364.

Liu, D., Diorio, J., Tannenbaum, B., Caldji, C., Francis, D., Freedman, A., Sharma, S., Pearson, D., Plotsky, P. M., & Meaney, M. J. (1997). Maternal care, hippocampal glucocorticoid receptors, and hypothalamic-pituitary-adrenal responses to stress. *Science, 277,* 1659–1662.

Liu, D., Caldji, C., Sharma, S., Plotsky, P. M., & Meaney, M. J. (2000). The effects of early life events on in vivo release of norepinephrine in the paraventricular nucleus of the hypothalamus and hypothalamic-pituitary-adrenal responses during stress. *Journal of Neuroendocrinology, 12,* 5–12.

Lomanowska, A. M., Boivin, M., Hertzman, C., & Fleming, A. S. (2015). Parenting begets parenting: a neurobiological perspective on early adversity and the transmission of parenting styles across generations. *Neuroscience,* http://www.sciencedirect.com/science/article/pii/S0306452215008489.

Lopez-Duran, N. L., Kovacs, M., & George, C. J. (2009). Hypothalamic-pituitary-adrenal axis in depressed children and adolescents: a meta-analysis. *Psychoneuroendocrinology, 34,* 1271–1283.

Lubin, F. D., Roth, T. L., & Sweatt, J. D. (2008). Epigenetic regulation of bdnf gene transcription in the consolidation of fear memory. *Journal of Neuroscience, 28,* 10576–10586.

Lupien, S. J., McEwen, B. S., Gunnar, M. R., & Heim, C. (2009). Effects of stress throughout the lifespan on the brain, behavior and cognition. *Nature Reviews in Neurosciences, 10,* 434–445.

Mastripieri, D. (1998). Parenting styles of abusive mothers in group-living rhesus macaques. *Animal Behavior, 55,* 1–11.

McEwen, B. C. (1999). Stress and hippocampal plasticity. *Annual Reviews in Neuroscience, 22,* 105–122.

McEwen, B. C. (2008). Understanding the potency of stressful early life experiences on brain and body function. *Metabolism, 57,* S11–S15.

McEwen, B. S., & Gianaros, P. J. (2010). Central role of the brain in stress and adaptation: links to socioeconomic status, health and disease. *Annals of the New York Academy of Sciences, 1186,* 190–222.

McGowan, P. O., & Szyf, M. (2010). The epigenetics of social adversity in early life: the implications for mental health outcomes. *Neurobiological Research, 39,* 66–72.

McGowan, P. O., Sasaki, A., D'Alessio, A. C., Dymov, S., Labonté, B., Szyf, M., Turecki, G., & Meaney, M. J. (2009). Epigenetic regulation of the glucocorticoid receptor in human brain associates with childhood abuse. *Nature Neurosciences, 12,* 342–348.

McKim, D. B., Niraula, A., Tarr, A. J., Wohleb, E. S., Sheridan, J. F., & Godbout, J. P. (2016). Neuroinflammation dynamics underlie memory impairments after repeated social defeat. *Journal of Neurosciences, 36*(9), 2590–2604.

Meaney, M. J. (2001). Maternal care, gene expression, and the transmission of individual differences in stress reactivity across generations. *Annual Reviews in Neurosciences, 24,* 1161–1192.

Meaney, M. J., Szyf, M., & Seckl, J. R. (2007). Epigenetic mechanisms of perinatal programming of hypothalamic-pituitary-adrenal function and health. *Trends in Molecular Medicine, 13,* 269–277.

Miller, G. E., & Chen, E. (2007). Unfavorable socioeconomic conditions in the early life presage expression of proinflammatory phenotype in adolescence. *Psychosomatic Medicine, 69,* 402–409.

Miller, C. A., & Sweatt, J. D. (2007). Covalent modification of DNA regulates memory formation. *Neuron, 53*, 857–869.

Miller, G. E., Chen, E., Fok, A. K., Walter, H., Lim, A., Nicholls, E. F., Cole, S., & Kobor, M. S. (2009). Low early-life social class leaves a biological residue manifested by decreased glucocorticoid and decreased proinflammatory signaling. *Proceeding of National Academy of Sciences of USA, 106*, 14716–14721.

Miller, C. A., Gavin, C. F., White, J. A., Parrish, R. R., Honasoge, A., Yancey, C. R., Rivera, I. M., Rubio, M. D., Rumbaugh, G., & Sweatt, J. D. (2010). Cortical DNA methylation maintains remote memory. *Nature Neuroscience, 13*, 664–666.

Mitchell, C., Schneper, L. M., & Notterman, D. A. (2015). DNA methylation, early life environment, and health outcomes. *Pediatric Research, 79*, 212–219.

Monk, C., Spicer, J., & Champagne, F. A. (2012). Linking prenatal maternal adversity to developmental outcomes in infants: the role of epigenetic pathways. *Developmental Psychopathology, 24*, 1361–1376.

Morgan, C. P., & Bale, T. L. (2011). Early prenatal stress epigenetically programs dysmasculinization in second-generation offspring via the paternal lineage. *Journal of Neurosciences, 31*, 11748–11755.

Mueller, B. R., & Bale, T. L. (2008). Sex-specific programming of offspring emotionality after stress early in pregnancy. *Journal of Neurosciences, 28*, 9055–9065.

Naumova, O., Lee, M., Koposov, R., Szyf, M., Dozier, M., & Grigorenko, E. L. (2012). Differential patterns of whole-genome DNA methylation in institutionalized children and children raised by their biological parents. *Development and Psychopathology, 24*, 143–155.

Nestler, E. J., Pena, C. J., Kundakovic, M., Mitchell, A., & Akbarian, S. (2015). Epigenetic basis of mental illness. *Neuroscientist, 22*(5), 447–463.

Oberlander, T., Weinberg, J., Papsdorf, M., Grunau, R., Misri, S., & Devlin, A. M. (2008). Prenatal exposure to maternal depression and methylation of human glucocorticoid receptor gene (*NR3C1*) in newborns. *Epigenetics, 3*, 97–106.

Pan, P., Fleming, A. S., Lawson, D., Jenkins, J. M., & McGowan, P. O. (2014). Within- and between-litter maternal care alter behavior and gene regulation in female offspring. *Behavioral Neurosciences, 128*, 736–748.

Pederson, D. R., Gleason, K. E., Moran, G., & Sandi, B. (1998). Maternal attachment representations maternal sensitivity and the infant-mother attachment relationship. *Developmental Psychology, 34*, 925–933.

Plotsky, P. M., & Meaney, M. J. (1993). Early, postnatal experience alters hypothalamic corticotropin-releasing factor (CRF) mRNA, median eminence CRF content and stress induced release in adult rats. *Molecular Brain Research, 18*, 195–200.

Provencal, N., & Binder, E. B. (2015). The neurobiological effects of stress as contributors to psychiatric disorders: focus on epigenetics. *Current Opinion in Neurobiology, 30*, 31–37.

Radtke, K. M., Schauer, M., Gunter, H. M., Ruf-Leuschner, M., Sill, J., Meyer, A., & Elbert, T. (2015). Epigenetic modification of the glucocorticoid receptor gene are associated with the vulnerability to psychopathology in childhood maltreatment. *Translational Psychiatry, 5*, e571.

Reynolds, R. M., Labad, J., Buss, C., Ghaemmaghami, P., & Raikkonen, K. (2013). Transmitting biological effects of stress in utero: implication for mother and offspring. *Psychoneuroendocrinology, 38*, 1843–1849.

Reznikov, A. G., Nosenko, N. D., Tarasenko, L. V., Sinitsyn, P. V., & Polyakova, L. I. (2001). Early and long-term neuroendocrine effects of prenatal stress in male and female rats. *Neuroscience and Behavioral Physiology, 31*, 1–5.

Roth, T. L. (2013). Epigenetic mechanisms in the development of behavior: advances, challenges, and future promises of a new field. *Developmental Psychopathology, 402*, 1279–1291.

Roth, T. L., Zoladz, P. R., Sweatt, J. D., & Diamond, D. M. (2011). Epigenetic modification of hippocampal bdnf DNA in adult rats in an animal model of post-traumatic stress disorder. *Journal of Psychiatric Research, 45*, 919–926.

Rozanov, V. A. (2013). Epigenetics: stress and behavior. *Neurophysiology, 44*, 332–350.

Sapolsky, R. (2015). Stress and the brain: individual variability and inverted U-curve. *Nature Neuroscience, 18*, 1344–1346.

Sasaki, A., de Vega, W. C., & McGowan, P. O. (2013). Biological embedding in mental health: an epigenomic perspective. *Biochemistry and Cellular Biology, 91*, 14–21.

Schroeder, M., Krebs, M. O., Bleich, S., & Frieling, H. (2010). Epigenetics and depression: current challenges and new therapeutic options. *Current opinions in Psychiatry, 23*, 588–592.

Stankiewicz, A. M., Swiergiel, A. H., & Lisowski, P. (2013). Epigenetics of stress adaptation of the brain. *Brain Research Bulletin, 98*, 76–92.

Szyf, M., Weaver, I. C., Champagne, F. A., Diorio, J., & Meaney, M. J. (2005). Maternal programming of steroid receptor expression and phenotype through DNA methylation in the rat. *Frontiers in Neuroendocrinology, 26*, 139–162.

Tyrka, A. R., Price, L. H., Marsit, C., Walters, O. C., & Carpenter, L. L. (2012). Childhood adversity and epigenetic modulation of the leucocyte glucocorticoid receptor: preliminary findings in healthy adults. *PLOS One, 7*, e30148.

Uchida, S., Hara, K., Kobayashi, A., Otsuki, K., Yamagata, H., Hobara, T., Suzuki, T., Miyata, N., & Watanabe, Y. (2011). Epigenetic status of Gdnf in the ventral striatum determines susceptibility and adaptation to daily stressful events. *Neuron, 69*, 359–372.

Unternaehrer, E., Luers, P., Mill, J., Dempster, E., Meyer, A. H., Staehli, S., Lieb, R., Hellhammer, D. H., & Meinlschmidt, G. (2012). Dynamic changes in DNA methylation of stress-associated genes (oxtr, bdnf) after acute psychosocial stress. *Translational Psychiatry, 2*, e150.

Vaiserman, A. M. (2015). Epigenetic programming by early life stress: evidence from human populations. *Developmental Dynamics, 244*, 254–265.

Weaver, I. C. G., Cervoni, N., Champagne, F. A., D'Alessio, A. C., Sharma, S., Seckl, J. R., Dymov, S., Szyf, M., & Meaney, M. J. (2004). Epigenetic programming by maternal behavior. *Nature Neuroscience, 7*, 847–854.

Weinstock, M. (2005). The potential influence of maternal stress hormones on development and mental health of the offspring. *Brain Behavior and Immunology, 19*, 296–308.

Weinstock, M. (2008). The long-term behavioral consequences of prenatal stress. *Neurosciences and Biobehavioral Research, 32*, 1073–1086.

Welberg, L. A. M., Seckl, J. R., & Holmes, M. C. (2001). Prenatal glucocorticoid programming of brain corticosteroid receptors and corticotropin-releasing hormone: possible implication for behavior. *Neuroscience, 104*(1), 71–79.

Wilkinson, R., & Marmot, M. (Eds.). (1998). *Social determinants of health*. Europe: WHO.

Zoladz, P. R., Conrad, C. D., Fleshner, M., & Diamond, D. M. (2008). Acute episodes of predator exposure in conjunction with chronic social instability as an animal model of post-traumatic stress disorder. *Stress, 4*, 259–281.

5

Interactions and Integrations— Biobehavioral Model of Suicide Based on Genetics, Epigenetics, and Behavioral Adjustment

Suicide is a complex multifaceted and multilayered phenomenon, which cannot be explained in simple and easy terms. It is serious public health problem throughout the world and is, thus, recognized in the modern society as a public health priority. Global suicide rates are growing steadily for the last 70 years, in the past 45 years, it has increased by 60%. Even though many countries report a downward trend in the suicide rate between 1981 and 2000, there has been a steady increase since then. For instance, in the USA since the beginning of a new century suicide rates have grown up to 40% in Native Americans and in middle-aged Whites, especially (and surprisingly) in women. While suicide is ranked as 10th leading cause of death in general, it is stated as the third leading death for children (aged 10–14) and the second leading cause of death among adolescents (aged 12–18) and young adults (aged 19–24) throughout the world, regardless of country gross income (WHO, 2010, 2013, 2014a, 2014b). In many countries, the tendency of growing suicides in the youngest adolescents is registered while older age groups demonstrate a decreasing trend.

Suicidologists around the globe keep calling attention not only to growing suicide rates but to this alarming "rejuvenation" of suicide. Several decades ago, suicide rates were not even considered for adolescents aged 14–15, while nowadays we are forced to face data for 10-year-olds and younger, even though these rates are low and unstable (McKeown, Cuffe, & Schulz, 2006, Rozanov, 2014). Adolescence is stated as a time of onset of suicidal ideation, attempts and completed suicides in all countries that

Stress and Epigenetics in Suicide. http://dx.doi.org/10.1016/B978-0-12-805199-3.00005-1

collect data on suicidal activity. Our analysis has confirmed that suicides in adolescents is a growing problem worldwide and that most serious rise is registered in developing countries or in the countries in transition, and in indigenous people in economically developed countries, such as USA, Canada, Australia, New Zealand, Sweden, Russian Federation, and others. Among the reasons, the pressure of the western culture may be named, which provokes consumerism, individualism, competitiveness, hedonism, an evolution of values and growing problems of mental health in children and adolescents. It is associated with enhanced psychosocial stress, which is dependent on inequalities, loneliness, family crisis, and lack of social support. Globalization and universal development of information technologies make these influences even stronger by involving bigger contingents and enhancing exposure to information about suicide (see Chapter 1).

Recent decades also were marked with extraordinary fast development in understanding neurobiological mechanisms of stress and delineating positive and harmful effects of the stress hormones on the organism and on the brain in particular. Together with the discovery of neuroplasticity that adopted new views on the brain as not only the central, but also as constantly changing organ, it created new thinking in neurosciences. Understanding of the role of stress, especially of early life stress in compromising critically important brain structures, such as the hippocampus, amygdala, and prefrontal cortex during brain development, promoted progress in understanding the neurobiological basis of mental health problems. Another great achievement was revealing of the contribution of epigenetic transformations in the development of stress-reactivity and in the genesis of neuropsychiatric disorders and disturbed behaviors.

On this background, new opportunities opened for the elaboration of integrative models of suicide which may encounter biological, psychological, and social factors in their complex interactions and interrelations. Though suicide is understood as a multifaceted phenomenon, medical models that consider suicide a complication of the mental illness are still prevailing, especially in suicide prevention. Growing prevalence of mental disorders and especially anxiety disorder and depression, which is another tendency for the last decades, may really have contributed to recent suicide rise in younger people. On the other hand, interest toward integrated models is not limited to finding links between mental health problems and decision to take one's life or impulsive gesture which may lead to suicide. Many suicides are committed by young people without any clear evidence of a mental disorder, so other factors should be taken into consideration. They may include behavioral patterns, especially risky behaviors, the contagion of suicides among youngsters and the role of suicide exposure, as well as higher rank factors, such as values, meaning in life, life purposes and goals, and perception of the future.

On the other hand, all these factors are closely interconnected. For instance, behaviors and feelings are influenced by negative life events, while frustrations, anxiety about future, and depressive thoughts constitute "internal" perceived stress. Internal feelings and emotions may also trigger biological mechanisms of stress and, as it may be hypothesized, may induce epigenetic transformations. Pathological pathway of such kind is well expressed in the title of one of the recent publications in this field—"Don't worry; be aware of the epigenetics of anxiety" (Nieto, Patriquin, Nielsen, & Kosten, 2016). The essence is that in animal models of anxiety epigenetic marks are found in limbic and cortical brain regions known to be involved in emotional responses to stress. Of course, animal models should not be translated to humans immediately, but it is well established that early life stressful events and adversities in childhood may be the reason of impaired brain biology and enhanced programmed stress-reactivity and that epigenetics is definitely involved in it. On the other hand, the evidence is accumulating, that later in life social events and interactions, for instance, those associated with socioeconomic status, can also leave epigenetic marks and change genes activity. This is a very strong support in favour of existence of reciprocal and recurrent relations between external events, internal emotions, epigenetic marks, and internally perceived stress. Of course, there is much more interest toward this topic so far as many registered stress-related epigenetic transformations produce phenotypes, which are transmitted to subsequent generations, though the nature of stressor is different from pure psychosocial. All this makes a background toward which we have undertaken an attempt to build an integrative model of suicide that combines biological, social, and psychological factors, a model based on the whole set of interactions and interrelations between these factors. One of the first necessary steps is to discuss main existing theories of suicide from the point of view how stress is integrated into these theories.

THEORIES AND MODELS OF SUICIDAL BEHAVIOR— HOW THEY ENCOUNTER STRESS

Suicide can be studied on different levels—on the population level and on the individual level. On the population level, suicide is seen as a statistically stable sociological phenomenon depending on many factors, among which are socioeconomic variables, for instance, social integration and social regulation according to Durkheim, prevalence of mental health problems, including addictions, genetic structure of the population, cultural, political, ideological and religious factors, as well as the level of life stress (Fig. 5.1).

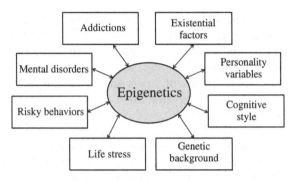

FIGURE 5.1 **Factors associated with suicide and possible central role of epigenetics** (*dotted lines*—**supposed links**).

Within this approach, studies focus on various biological, social, and personality (psychological) risk factors of suicide and try to evaluate how these factors interact, moderate, modulate, or mediate each other while influencing suicidal outcomes. On the other hand, the growing number of identified and interacting risk factors results in very complex and blurred picture that does not add much to the prevention methods. Moreover, risk factors are too much dependent on cultural aspects and due to it often contradict each other, for instance, in the western culture being male means higher suicide risk, while in China and in some other regions in Asia, young women commit suicide more often than men. Occupational risk factors are also not the same in the western or eastern societies, etc.

On the individual level, attention is concentrated on most common pathways that may lead to suicide. Within this approach, such factors as psychosocial variables, existential problems, beliefs and experiences, individual stress-vulnerability, and severity of depressive symptoms may have most serious importance. The individual combination of psychological traits, such as impulsivity, neuroticism, extraversion/introversion, as well as cognitive rigidity and pessimism should be taken into consideration. Many other circumstances may also affect individual risk, for instance, suicide exposure (being familiar or impressed by suicide of a significant other), personal values, beliefs, and attitudes toward life and death, etc. Though suicide completers are a much differentiated group, many authors have tried to find commonalities that may characterize this group and to build a generalizing theory of suicide. Nevertheless, there is no one single theory of suicide that would comprise all the known generalities of why suicides happen, and possibly will never appear, given the complexity of the phenomenon. Even the causes of suicidal activity are not fully comprehended till now and all schematizing regarding risk factors, individual vulnerability, and behavioral outcomes should be regarded as probabilistic, not deterministic (van Heeringen, Hawton, & Williams, 2000; O'Connor & Nock, 2014).

Recently O'Connor & Knock have given an analysis of the most influential suicide theories and models and came to a conclusion that most modern theories are "diathesis stress in origin and cognitive in focus," while earlier elaborated theories emphasized predominantly individual psychological factors (O'Connor & Nock, 2014). Indeed, after 1970s the focal point of explanation why people die from suicide moved to either some cognitive judgment deficits (represented by R. Baumsteiner, E. Schneidman, M. Rudd, J. Williams), or vulnerability acquired in the childhood or caused by prolonged stress (J. Mann, D. Wasserman, A. Beck, T. Joiner, R. O'Connor). The biggest progress in understanding suicide recently is expressed in terms of neurobiological mechanisms of suicidal behavior and interaction among biological, social, and personality factors. In the majority of theories, stress possesses one of the central positions.

It is impossible to characterize all existing theories of suicide here, but it may be interesting from the perspective of our discussion to compare how different approaches have evaluated stress in the row of other factors of suicide. It may be said that stress exists in every known theory (maybe with the exception of Durkheim's sociological etude and Freud's libido and mortido opposition, so far as the concept of stress was not known at that time). In some theories stress is just taken into account as a factor, while some theories are fully grounded on the concept of stress, and particularly such traits as stress-vulnerability or, as it is often conceptualized, stress-diathesis. We will here comment on the most important and well-known ones.

Durkheim's sociological theory was based on such broad factors as social and moral integration and regulation of the society. Though it was discussing suicide in the row of other deviant behaviors and did not directly touch psychological variables, it had an impact on all subsequent theories including pure psychological ones (Rudd, Trotter, & Williams, 2009). Durkheim (1897) only seemingly did not intrude into personality and intrapsychic experiences, his explanations of the typology of suicide gives much for understanding individual suicidality in the social context. Further, many authors have taken elements from Durkheim's theory, such as economic situation and possibility of individuals to meet main survival needs (housing, food, clothing, etc.), as well as other elements, like self-identification and self-image (Rudd et al., 2009). Farber (1968) in his theory tried to explain suicide rate as the function of the percentage of highly vulnerable individuals and the level of adversities affecting the given population, pointing that the less acceptable are life conditions, the less hope it is in society and the higher is suicide potential (Farber, 1968). Such theory may be seen as a direct continuation and development of Durkheim's views but with introducing such variable as individual vulnerability. On the other hand, it lacks such components as collective feelings or spirits in the society that may help to tolerate the utmost level of adversities and in this sense concedes to Durkheim's views.

Several theories are focusing on internal experiences of the suicidal personality and cognitive processes. While R. Baumeister focused on the idea of suicide as an escape from self (Baumeister, 1990), E. Schneidman outlined 10 commonalities of suicide and developed a concept of psychache (Schneidman, 2001). Though stress as itself is not the central factor within these cognitive theories, it is interesting to mention, that Baumeister's theory included such essential components as frustrated goals and personal failure followed by self-blame. Within this concept, individual is escaping not only from the self, but also from internal negative affect. There is rather a clear similarity between these two theories so far as according to Schneidman suicidal person is also deeply frustrated and is trying to cease consciousness because of unbearable psychache (Rudd et al., 2009). With regards to the concept of psychological or emotional stress, which is characterized by negative thoughts and frustrations, inability to achieve one's goals and rumination—everything that is mentioned in these theories as internal psychological torments, can be referred as severe perceived (or subjective) emotional psychological stress.

One of the influential psychological theories of suicide is Beck's cognitive theory, in which central role possesses the feeling of hopelessness and pessimism about future (Beck, Brown, Berchick, Stewart, & Steer, 1990). Beck and his colleagues speculated that pessimism of the suicidal person is overwhelming and captures beliefs about self, others, and future (Rudd et al., 2009). He also developed cognitive theory and proposed the existence of "modes"—cognitive, motivational, and behavioral schemas that are able, if constantly repeated, to create vulnerability to future problems. Thus, Beck's cognitive theory can be referred as a stress-diathesis model in which dispositional vulnerability plays an important role (O'Connor & Nock, 2014).

Rudd developed the theory of modes further and conceptualized the "suicidal mode" associated with hopelessness and described it as a "suicidal belief system." Rudd developed a "fluid vulnerability theory," a more dynamic and interactional understanding of suicidality (Rudd, 2006). In this cognitive-behavioral theory predisposing vulnerabilities, associated with suicidal mode (which includes developmental trauma and parental modeling) interact with triggers or stressors—internal (thoughts, feelings, physical sensations) and external (situations, circumstances, people, places, etc.). The central role of internal stress in this theory is remarkable. These interactions are linked to the cognitive system, represented by suicidal beliefs, which in turn, if supported by affective and physiological systems and ultimately leads to activation of the behavioral system, directly related to preparations to suicide and suicidal act (Rudd, 2000).

Though this and previous theories are labeled as cognitive, they operate with concepts, which are fairly coinciding with what is understood as stress-vulnerability.

This vulnerability or higher sensitivity to stress is most often seen as a predisposing trait that is acquired in the early childhood, or due to repeated stressful (including internal psychological) situations. Williams in his over-general memory and the "cry of pain" model concentrates on cognitive mechanisms. The central point is that suicidal person, particularly one who is depressed or has a history of suicide attempt, often performs poorly when required to produce a specific memory in response to a cue word, remembering things in summarized form (Williams & Pollock, 2000). It impairs problem-solving process, when instead of recalling specific strategies while being challenged, individuals may react with despair, which limits ability not only to adapt but also to think in future terms. Williams has incorporated his theory into "cry of pain" model of suicide associated with the feeling of entrapment, which has similarity with the ethological construct of "being trapped" (Goldney, 2000). It may be noticed that findings regarding overgeneral memory are in a very good agreement with modern knowledge of compromised hippocampal structures in stressed individuals. Feeling of being trapped without the chance to escape is an analog of a social fear induced by past humiliations, including the feeling of defeat and being humiliated in life (Rudd et al., 2009). One can find here links to escape theory and speculate that these feelings and emotions may reflect severe internal stress, ultimately leading to a perception that suicide remains the only choice.

One of the popular modern theories is an interpersonal-interactional theory of T. Joiner. This concept is putting forward three important factors, which may lead to suicide: (1) feelings that one does not belong with other people (thwarted belongingness); (2) feelings that one is a burden on others or society (perceived burdensomeness); and (3) an acquired capability to overcome the fear and pain associated with suicide (acquired capability of lethal self-injury) (Joiner, Brown, & Wingate, 2005). The first two interpersonal constructs are postulated to promote a desire to die, while capability for suicide is the function of repeated stressful and painful experiences, such as traumas, self-injury, prior attempts, or exposure to physical fight and pain. Development of ability for lethal self-injury may be also enhanced by persistent suicidal thoughts. Within the frame of this theory suicides in some specific groups may be understood better. For instance, war veterans with combat experiences, when returning to civil life, may go through such events as losing relations with their comrades and feeling excluded from civilian life, especially if their home situation is not satisfactory and friendly (in terms of relations, finances, unemployment, crisis, etc.) (Rozanov & Carli, 2012). On the other hand, combat training and military exposures, staying in a situation of constant tension and vigilance, being exposed to deaths around, pictures of dead bodies, being trauma-tized or injured may cause habituation to painful experiences, which re-ally may mean lower threshold for suicide (Selby et al., 2010). In view of this, it is interesting to mention hypothesis of I. Orbach that at least some suicidal individuals are characterized by relative insensitivity to physical pain and indifference to their bodies. The reason for higher tolerance to pain (confirmed empirically) may be early and continuous stress that may lead to the development of dissociative tendencies in the form of indiffer-ence to the body and pain and heightened vulnerability to stress. These dispositions may facilitate suicidal behavior in the face of mounting in-tolerable stress, helplessness, and hopelessness (Orbach, 1994). Moreover, there are some other peculiarities in suicidal depressed persons, like elec-trodermal hyporeactivity, suggesting some basic physiological abnormali-ties that may be related both to stress, depression and cognitive processes (Thorell, 2009).

Thus, practically all theories of suicide which are referred as "psycho-logical" are actually either indirectly speaking about psychological stress, or directly include stress-vulnerability as a dispositional factor and incor-porate stressful experiences as triggering events. Many internal cognitive processes, especially hopelessness, pessimism, feelings of being trapped, frustrations, negative affect, negative self-perception and others are noth-ing but signs of psychological or emotional stress, which include thoughts and emotions that may become stressful per se in the situation of chro-nicity. All previously mentioned theories are psychological in nature and with prevailing cognitive component, thus neuroendocrine mechanisms of stress are not at the center of discussion but are used as a confirmation or biological proxies of psychological disturbances.

There are also several very well-known and famous theories of suicide, like the psychodynamic concept of Z. Freud and construct of the "wish to die, wish to kill, and wish to be a victim" of K. Menninger that quite well do without representations of stress. There are also philosophical views that discuss quite different basic issues of the human consciousness and that propose their own explanations. On the other hand, there is another set of theories, which are neurobiological in nature and which are fully based on the concept of stress. They are exploring stress-vulnerability or stress-diathesis as a predisposition to suicide in the first turn and are sup-ported by clinical and neuroimaging data.

The stress-diathesis model explains suicide as the result of an interac-tion between environmental stressors and a trait-like diathesis or suscep-tibility to suicidal behavior (Mann, 1998; van Heeringen & Mann, 2014). Thus, stress diathesis is seen as a trait or a quality that makes suicide in one person more probable that in another while both are experiencing the same level of stress (for instance, the same number and severity of negative life events). So far as there is a strong subjective component in a perception of negative life events, the severity of disturbances in brain systems and circuits that take part in stress evaluation and cognitive reappraisal, as well as in fear, anxiety, and anger are crucial. The strong side of the stress-diathesis model is that it is supported by findings from post-mortem studies of the brain and from genomic and in vivo neuroimaging studies. Such studies of stress, as we already have pictured in details in previous chapters, usually indicate impairments of the serotonin neurotransmitter system in the prefrontal cortex, hippocampus shrinkage, amygdala hypertrophy and the HPA stress-response system regulation abnormalities. These impairments are logically associated with deficits in control of mood, pessimism, reactive aggressive traits, impaired problem solving, over-reactivity to negative social signals, excessive emotional pain, and suicidal ideation (van Heeringen & Mann, 2014).

Stress-diathesis model is an example of the attempt to understand suicidal act on the basis of the central role of stress, which is quite logical and integral, though to a certain extent is one-sided. According to this model susceptibility to suicide is relatively independent of psychiatric disorders, and suicidal behavior is understood as an independent form of behavior. On the other hand, within this approach pathways leading to suicide are based mostly on neurobiological constructs that tend to explain all complexity of the suicidal process. One of the integrative approaches that are free from above-mentioned shortages and is more balanced with regards of the combination of biological, psychological, and social factors is the concept of D. Wasserman (Wasserman, 2001, 2016; Wasserman & Sokolowski, 2016). This model is also based on the idea of stress-vulnerability, but in addition, utilizes the concept of "suicidal process" and such combination make it different. The central idea of this model is that suicide is not a "bolt from the blue." Suicide is seen as a process which is unfolding in time, even if sometimes it looks as an unexpected event. The suicidal process according to D. Wasserman is associated with a variety of events, visible and invisible, including weariness of life, death wishes, suicidal thoughts of different severity, suicidal communications (suicidal messages), suicide attempts, and completed suicide (Wasserman, 2001, 2016). The model puts together several cognitive concepts, that is, conceptualization of suicidal behavior as suicidal ideation, attempts and completed suicide, a weariness of life concept with such components as vague death wishes and more definite suicidal thoughts and stress-diathesis model based on neurobiological findings. The suicidal process according to the model occurs in the context of two broad factors—risk factors, among which stress is the most important, and protective factors, which attenuate the negative effects of risk factors (Fig. 5.2).

With regards of counteraction of these factors, the suicidal process is represented as a series of alternating components of suicidal behavior, which can be either exacerbated or attenuated. For instance, suicide

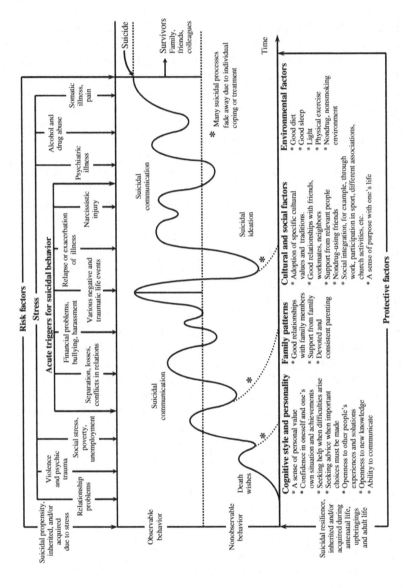

FIGURE 5.2 Stress-vulnerability model and development of the suicidal process from suicidal ideation to suicide. *Source: From Wasserman, D. (Ed.). (2001). Suicide. An unnecessary death, Martin Duniz; London; by permission.*

attempt may act either as a behavioral pattern that will help to acquire capability for repeated attempts and enhance suicide risk, or a sort of emotional discharge event, that may result in catharsis. Very much will, of course, depend on the balance between further risk and protective factors.

Stress is represented in the form of stress-vulnerability (predisposition due to genetic make-up of the personality or enhanced due to early life adversities) and a variety of chronic and acute stressors, that reflect major problems of the personality (relationship problems, violence, social stress, mental or somatic illness) and immediate triggers of suicidal behavior (negative life events, financial problems, separation and losses, conflicts, life failures, etc.). Thus, within this concept stress-vulnerability, which is understood primarily as genetic predisposition, may be enhanced due to early life stress which leads to vulnerable personality type and higher stress-reactivity, and which is being tested throughout life by major negative life circumstances and triggers of suicidal behavior. These stressful experiences interact with invisible and visible suicidal activity and protective factors. Protective factors are an important part of the scheme so far as their list itself is a resource for conceptualizing prevention strategies and planning of practical measures of prevention (Wasserman, 2001, 2016; Wasserman & Sokolowski, 2016).

Summarizing this discussion, it may be said, that besides very early theories of suicide stress as a factor of suicide is present practically in all influential theories, both those that are referred as psychological, with prevailing cognitive explanations and those which are fully based on the concept of stress. The last ones are dominated by neurobiological thinking and reasoning, the central idea of such theories is the stress-diathesis (or stress-vulnerability) which is understood as genetic predisposition. This predisposition, which is based on the genetic make-up of the individual may be enhanced by early life stress, and there is a high probability that epigenetic transformations are taking part in it. One of the most balanced models that combine biological, psychological, and social factors is the concept of D. Wasserman that represents suicidal process within the context of risk and protective factors. In the whole stress-diathesis models implies multiple strategies for their confirmation, that include a search of genetic markers of existing predisposition (genetic make-up of the individual) and environmentally induced epigenetic marks that lead to enhancement of stress-vulnerability, as well as direct signs of compromised brain systems, for instance, with the use of neuroimaging approaches.

HERITABILITY OF SUICIDE—CLASSICAL BEHAVIORAL GENETICS STUDIES

One of the popular myths about suicide, widely publicized in a variety of printed and electronic sources reads as follows: suicide is inherited. This statement gives birth to mental rationalization—prevention is very problematic, someone who is genetically predisposed will inevitably commit suicide. Usually after this follows myth debunking, as a rule, insufficiently grounded, a typical example of how in a simple phrase one tries to explain a rather complex phenomenon. At the same time, the problem of "heritability" of suicide is really complex and, as everything around suicide has no simple explanation.

From the point of view of behavioral genetics, it is more correct to speak about the contribution of genetic and environmental factors to this particular form of behavior. Behavioral genetics (sometimes also called psychogenetics) is a popular discipline and area of study devoted to studies of the differential contribution of genes and environment in human qualities, psychological characteristics, behavioral patterns, and psychopathologies (Gilger, 2000; Torgersen, 2005; Rozanov, 2013). Behavioral genetics deals mostly with complex polygenic phenotypes, usually represented in the population by a normal distribution, like most of the psychological traits or psychophysiological variables. From the point of view of behavioral genetics family history of suicide may be determined either by shared genes or shared family environment, or both. On the other hand, it is insufficient to estimate the contribution of genes and environment, it is much more important to understand the role of the interaction of genes and environment in shaping the suicidal behavior, which will be a focus of our discussion. Since the time of the creator of behavioral genetics (at that period referred as human genetics) Francis Galton, the central question was posed as a dichotomy—"nature or nurture," genetic factors or external (social, interpersonal, etc.) influences. Nowadays, though "nature of nurture" still remains a central question, suicidology is utilizing more integrated bio-psychosocial approach, considering the pathogenesis of mental and behavioral disorders, as well as suicidality in terms of the constant interaction of genetic, psychological and social factors (Engel, 1997; Moffitt, 2005; Pilgrim, 2002; Wasserman, Geijer, Sokolowski, Rozanov, & Wasserman, 2007b).

Methodologically behavioral genetics is based on three main approaches—family, twin, and adoption studies. Suicidality as a behavioral pattern, no doubt, runs in families. This fact has long been observed in human history and has been confirmed by many studies which were repeatedly overviewed (Turecki, 2001; Brent & Mann, 2005; Brent & Melhem, 2008). For instance, patients referred to the suicide prevention centers and psychiatric patients hospitalized after suicide attempts often report family history of completed suicide and attempted suicide (Robins, Schmidt, & O'Neal, 1957; Murphy, Wetzell, Swallow, & McClure, 1969). Psychiatric patients, who have a family history of suicide, are more likely to commit suicide attempts than patients with mental disorders but without family history of suicide, and vice versa, the relatives of patients with a history of suicide attempts have higher frequency of attempts, than relatives of patients, who never had a history of suicide attempts (Roy, 1983). Having a family history of suicide significantly increases the risk of suicide attempts and completed suicide in patients with various forms of affective disorders. Approximately 10% of patients with depression have a family history of suicide (Roy, 1993). From various studies, it was concluded, that the average risk of suicide among relatives in suicidal families is about 5 times higher than in nonsuicidal, and the relative risk is higher in the case of completed suicide compared with the attempts (Baldessarini & Hennen, 2004).

However, a family method of behavioral genetics has many limitations. The fact that suicide runs in families do testify of the presence of genetic influences on suicidal behavior, but does not give a possibility to evaluate differential input of genes and environment, so far as along with the decrease of the percentage of common genes in different grade relatives, the fall of the influence of the shared environment also takes place. One of the most serious questions that remain open in family studies is whether we are dealing with a biological or social/psychological heritability, when suicide history in the family reduces the threshold for such behavior in other relatives, especially younger ones. Psychological theories of suicide promote the role of social learning and behavioral transmission of suicide. For instance, from the psychodynamic perspective family secrets, stories (narratives) and other manifestation of the "family unconscious" may play a crucial role in the formation of suicidal tendencies in future generations (Schützenberger, 1998). Another question that rises from family studies, is whether suicidality is inherited as an independent behavioral pattern or whether it actually follows heritable (also partly, with a significant influence of the environment) psychopathologies or mental illnesses, given that presence of mental disorder, especially depression, significantly increases the risk of suicide. A number of studies have been focused on the elucidation of these questions.

One line of evidence was based on the assumption, that if transmission of suicidality is predominantly sociopsychological and even unconscious, then one would expect that it will manifest itself at certain stages of the life cycle, for instance, during adolescence or early adulthood, when examples from the older generation have bigger impact on emerging patterns of behavior in the young. Several studies have proved that it is not the case and that family aggregation of suicide can be found both in young and old age with almost the same quantitative pattern (Batchelor & Napier, 1953; Garfinkel, Froese, & Hood, 1982; Murphy & Wetzel, 1982). This rather indirect evidence was later supported by studies involving other methodological approaches, like twin and adoption methods and that will be discussed further.

Of much bigger importance was the problem of suicide as an independent or linked to psychopathologies behavior, therefore a lot of attention was paid to the coincidence of suicidal behavior and mental disorders (especially depression or manic-depressive disorder) on the family level. Family suicidality is really often associated with a family history of mental illness, but as it was showed in many studies, suicide may be associated with different mental illnesses implying independent heritability of suicidal behavior. Thus, A. Roy has followed more than 11,000 psychiatric patients; from them 5845 patients had a family history of suicide and 5602 had not. Suicide among first and second generation biological parents increased the risk of suicide attempt in their offspring in the presence of a variety of mental disorders (schizophrenia, unipolar and bipolar depression, neurotic states, personality disorders, alcohol addiction) (Roy, 1983; Roy, Rylander, & Sarchiapone, 1997).

Another well-known large-scale study of the same type is IOWA-500. In this study, histories of lives of more than 500 patients of a psychiatric clinic in Iowa diagnosed with schizophrenia, manic-depressive, and depressive disorders and their relatives and descendants were traced for 30–40 year. As a control, a group of surgical patients was used. Totally 30 cases of suicide were found—29 among the 525 psychiatric patients and 1 out of 160 people in the control group. It was also shown that the presence of mental disorder (particularly depression) is combined with an enhanced (about 4 times) suicide risk in children of psychiatric patients when compared with children of patients of the control group, among other relatives of psychiatric patients suicide was observed 3 times more often than in the control group (Tsuang, 1983). In this study, it was confirmed that suicide may occur in the presence of different psychiatric disorders.

Another widely cited research relevant for understanding of suicide as an independently inherited form of behavior is the study of the Amish community in the United States. This study provided a unique opportunity to trace suicidality at the family level in highly controlled conditions. Amish Community (Lancaster County, southeast of Pennsylvania) is a group of compactly residing Anabaptists, ethnically homogeneous pacifists, and living very simple life, reluctant to adapt technological novelties. The community is practically alcohol-free, characterized by the absence of any forms of violence, and aggression (homicides in the community are unknown from the beginning of its existence). The Amish community is associated with strong religious beliefs and is known for high level of social integration and mutual support, divorces, and family problems are extremely rare, economically they are rather prosperous (all of them are wealthy farmers). Thus, in the Amish community many risk factors for suicide practically do not exist. This community was studied by Egeland & Sussex in the 1980s of the last century. It was not surprising that cases of suicide were quite rare among community members—authors were able to identify a total of 26 completed suicides for the period of 100 years, from 1880 to 1980. It turned out that 24 of the 26 suicide victims had a history of affective disorders, and 73% of all diagnosed cases of suicide were reported in four primary pedigrees, containing also a heavy loading for affective disorders. This supports the idea of transmission of depression which is followed by suicide. But in the same community other family pedigrees have been identified with about the same frequency of affective disorders, but without suicides, suggesting that presence of affective disorders was not itself a predictor of suicidal behavior. The main conclusion, which was made on the basis of this study, is that the inheritance of suicidal behavior is relatively independent and may be not directly related to inheritance of affective disorders (Egeland & Sussex, 1985).

Further confirmation for this conclusion was received in a number of longitudinal studies. Analysis of 496 young people aged 10–21 years who had committed suicide during 1981–97 in Denmark and 24,800 controls matched for sex, age, and other variables showed that the strongest risk factor for suicide was mental illness in the young people, while effect of the parents' socioeconomic factors decreased after adjustment for a family history of mental illness and a family history of suicide (Agerbo, Nordentoft, & Mortensen, 2002). In another study, it was also found that completed suicide and psychiatric illness among relatives act as risk factors for suicide, but the effect of family suicide history is independent of the familial clustering of mental disorders (Qin, Agerbo, & Mortensen, 2002). On the other hand, one can agree with G. Turecki (2001) who points that family aggregation of affective disorders can interact with other family factors that predispose to suicide, such as personality traits, features of temperament, stress-vulnerability, which are also familiar. All these qualities, traits, and characteristics are polygenic and depend on multiple genes, each of them can make both positive and negative contribution to overall suicidality (Turecki, 2001). Moreover, many genes that are associated with suicide appeared to be important regulators of critical neurotransmitter systems (serotonergic, dopaminergic, GABA-ergic, etc.) or linked to brain cellular assembling (BDNF). All these genes due to their central role in neural tissue may produce nonspecific risks for a variety of psychiatric disorders and pathological behaviors, which raise a new set of questions about the factors that lead to more specified behavioral abnormalities and disorders. It may be also mentioned that existential factors and cognitive styles may have an independent impact on both mentally disturbed and intact individuals.

Another type of research, which enables more specific judgment regarding the contribution of genes to suicidal behavior, is twin studies. Twin studies constitute a traditional and very valuable approach in behavioral genetics. Actually, it is one of the first objective methods introduced by F. Galton more than a century ago. The method is based on the acknowledgement of differential dosage of identical genes in each type of twin pairs—monozygotic (MZT) twins share 100% of their genes, while dizygotic twins (DZT) only 50%, like siblings. On the other hand, in case of twins, one can rely on the unifying influence of the shared environment, especially during fetal and early postnatal period. In fact, things are much more complicated, because even in the MZT during prenatal development serious differences may occur due to differential level of circulation and nutrition, which may be associated with the epigenetic transformations, leading to differential genes expression and subsequent phenotypic differences. Besides, the level of unification of shared psychosocial environment in MZT and DZT pairs at different stages of identity formation is not the same and it may distort results. Nevertheless, if the risk of suicide is transmitted genetically, concordance should be higher among the MZT than in DZT.

Twin studies of suicidal behavior were systematically and on large samples performed by A. Roy. He showed that concordance rate of suicides among MZ twin pairs is really higher than among the DZT. In one of the studies, 176 twin pairs in which one cotwin had committed suicide were enrolled. In 9 of these pairs both cotwins have committed suicide, of these 7 pairs belonged to MZ twins (from total 63 monozygotic pairs), and only 2 belonged to dizygotic (from total 114). Revealed difference in the concordance rate is very convincing—7 out of 62 against 2 of 114 (Roy, Segal, Centerwall, & Robinette, 1991). When all known to date studies of this kind (which are not very numerous) are combined, the difference becomes even more pronounced—from 129 MZT pairs, there were 17 cases where both cotwins completed suicide, while from 270 DZT pairs—only 2 cases (Roy et al., 1997). A similar situation is observed in the case of suicide attempts. In the study of 35 survivor cotwins whose twin siblings have died of suicide, it was found that 10 from 26 survivors of MZ twin pairs have attempted suicide, while none of the survivors from DZ pairs attempted suicide (Roy, Segal, & Sarchiapore, 1995). Similar results were obtained by Statham et al. (1998) who studied a large sample (5995) of twins and revealed that in MZ twin pairs, as compared with DZ pairs, concordance rates were higher both for suicidal thoughts and serious suicide attempts. After statistical control of other risk factors, history of suicide attempts and suicidal thoughts remained stable and strong predictors of suicidality among MZ pairs and did not have a significant value in DZ twins. On the basis of this study authors have evaluated the contribution of genetic factors in suicidality, which appeared to be about 45% (Statham et al., 1998).

Twin studies are very conclusive, but also often criticized, especially when it comes to death of one of the cotwins. Twin studies with their formal approach may not encounter some psychological and interpersonal factors that are inherent to twins. It is known that between monozygotic twins in the process of personality formation unusually close attachment can emerge. Homozygous cotwins, due to their external similarity and possibly due to behavioral and emotional links during development in childhood, often start to identify themselves with their siblings, so that in their perception their cotwin becomes an inseparable part of themselves. From this, an extremely pronounced sense of loss and a special feeling of sorrow can emerge, which depicts deep attachment of surviving cotwin to the diseased. Such twins are known as "lonely twins." From their own perspective feeling of loss that they experience is unique and can be only understood by the same lone twins as they are (Woodward, 1998). It is clear that such psychological peculiarities may themselves be the cause of higher concordance for suicide or parasuicide in MZ twins, which can distort results of the studies.

Trying to avoid this complication, A. Roy hypothesized that if the death of one of the cotwins occurred not due to suicide, but to any other reason, the frequency of suicide attempts among lonely mono- and dizygotic twins should not differ significantly. He compared the frequency of suicide attempts among 166 MZ and 79 DZ twins, who lost their cotwin for reasons not related to suicide. It was found, that in this case, the difference in the frequency of suicide attempts was really negligible and even contradictory to possible genetic influence (1.8% among MZ and 3.7% among DZ twins) (Segal & Roy, 1995). Thus, twin studies have provided much more objective evidence for genetic basis of suicidal behavior.

Another line of conclusive evidence comes from the studies utilizing the third classical method of behavioral genetics—adopted children method. The power of such studies is that children, who for some reasons were separated from their families at birth or in early childhood, may be carrying genetic load of their biological parents, while all subsequent life events happen to them as part of their life in adoption families. In one of such studies, authors have used the registry of adopted children in Copenhagen which included 5483 adoptions for the period from 1924 to 1947. In this cohort analysis of the causes of death revealed 57 suicides. This group was matched with the control group of 57 adopted children, controlled for age, sex, social status of biological and adoptive parents, as well as the time they spent with their biological parents or in the institutions until entering their new families. Analysis of the causes of the death of their biological relatives revealed that from 269 parents of 57 identified adopted suicides 12 also committed suicide. At the same time, only 2 out of 269 parents of the control group committed suicide. No history of suicide or suicide attempts was registered in the families that have adopted individuals from both groups (Shulsinger, Kety, Rosenthal, & Wender, 1979). Adoption studies are considered the most conclusive in behavioral genetics. At the same time, this type of studies has always been the subject of criticism, mostly from the point of view of possible specific nature of the sample of biological parents. Recently, they have been criticized even more because peculiarities of in utero and early life experiences of adopted children remain largely unknown. On the other hand, it may have crucial importance, so far as new data pointed to the role of stress in utero and maternal behavior in programming mental health of offspring. It is, therefore, becoming clear that influences from biological parents may be much wider and very difficult to control, and this was not always taken into consideration in the studies cited earlier.

However, formal calculations from twin studies predict genetic input as 45%, while family studies estimate the risk of transmission of suicidal behavior from parents to children as 12–18%, among siblings—10–15% (Statham et al., 1998; Brent, Bridge, Johnson, & Connolly, 1996). A recent large-scale study (analysis of suicide cases among relatives of 83951 people who committed suicide in Sweden in the period from 1953 to 2003 in comparison with the control group) helped to clarify the risk of familial clustering of suicides (Tidemalm et al., 2011). Statistical analysis showed that the risk among siblings (OR = 3.1) was higher than that for maternal half-siblings (OR = 1.7), despite similar environmental exposure. It was confirmed that MZ twins are at higher risk than DZ twins, also it was shown, that the third-order relatives still are at higher risk as compared to controls. Thus, genetic factor can be traced even if the percentage of shared genes is rather low. On the other hand, shared (familial) environmental effects were also indicated; siblings to suicide decedents had a higher risk than their offspring (both 50% genetically identical, but siblings much more influenced by family shared environment, OR = 3.1 vs. 2.0), and maternal half-siblings had a higher risk than paternal half-siblings (both 50% genetically identical but the former more subjected to shared environment). This population-based study generally confirms all previous studies which are often criticized because of the insufficient number of observations or underestimation of the influencing factors.

Thus, classic behavioral genetics studies suggest that genes contribute significantly to suicidal behavior. On the other hand, such studies, in spite of their importance for understanding genetic and environmental input in the existing variability in the given population, have natural limitations. First, such studies have no potential to identify specific genes that are involved in the transmission of suicidal behavior; this can be done only on the basis of modern genotyping techniques and using the methodology of association or linkage studies. Other limitations are associated with psychological or existential factors that are prevailing around such sensitive phenomenon as suicide, including subconscious impulses. Moreover, some environmental influences cannot be encountered by controlling samples. Among the last are in-utero effects, early life parental influences, etc. It is especially actual today when mechanisms of epigenetic programming by parents' behavior and stressful influences are becoming clearer.

These transgenerational effects may have much more power than one could anticipate. All this reasoning reflects recent tendency to attach more importance to the environment than to genes. There is also another important direction of studies in this field—elucidating not only differential input of genes and environment but also interactions between these two factors. These interactions have been studied extensively and they really have a great potential to explain much of the complexity of inheritance of suicidal behavior.

SUICIDE GENETICS—GENES-TO-ENVIRONMENT INTERACTIONS AND VULNERABLE PHENOTYPES

The main task of behavioral genetics is not only to assess the differential role of genotype (inherited predisposition) and the environment (in the broadest sense—family, close associates, peers, social situations, general psychoemotional, cultural, and mental context) in the formation of the variability of psychopathology and suicidality, but also to evaluate the role of interaction of these two factors. This interaction can be estimated using classical psychogenetic methods and within approach, associated with the development of genomics, and possibilities of identification of gene polymorphisms. Fast progress in genotyping techniques, including genome-wide studies and even full genome sequencing (which is becoming cheaper as new automated techniques are developed) open new opportunities in identifying genetic background of suicidal behaviors. Both approaches are actively using advances in neuroscience, biological psychology, and psychiatry. There are two main variants of genes and environmental interaction: (1) genes–environment covariation (rGE) and (2) genes–environment interaction (GxE). In the first case, the analysis is aimed to evaluate how effects of genotypes and environments strengthen or weaken each other in a bilateral manner, in the second—how genotypes are moderating effects of the environment on behaviors or psychopathologies (Rowe, 2003). In the first case, interactional effects are associated mostly with behavioral aspects, while in the second case—with statistical correlations. In both cases interactions may be evaluated only on the population level, giving a general understanding of the tendency, while on the individual level these tendencies may be distorted and even reversed.

In its most general form genes–environment covariation (rGE) can be described as passive, reactive (evocative) and active. For instance, the so-called "poor parenting style" (a large set of environmental risks arising from adverse maternal or paternal relations, ranging from maternal smoking during pregnancy to inadequate responses to the behavioral peculiarities of the child) significantly predicts increased child aggressiveness, which include strikes, pushing, bullying, and other antisocial manifestations in relations with peers (Lahey, Moffitt, & Caspi, 2003). It demonstrates passive covariation of genes and environment, confirming intuitively clear and well-studied by clinical psychologists trend, when a bad climate in the family, due to psychological problems, or psychopathology of parents, is transmitted to children by social learning together with genes predisposing to mental health problems.

Reactive covariation can be demonstrated by the following observation: conflicting and negative parental behavior directed at one of the teen children in the family explains more than 60% of the variance of antisocial behavior and 37% of the symptoms of teen depression (Reiss et al., 1995). In another word, negative reactions of parents, irrespective of their parental feelings, to the child that exhibits unwanted behavior, in this or another way may enhance his/her psychopathological tendencies. Finally, active covariation of genes and environment reflects a nonrandom distribution of environments among different genotypes. In other words, "good" genotype is more likely during development to acquire by self-selection "good" environment, while "bad" genotype chooses the most adequate and "bad" environment. This can be illustrated by the study of Cleveland, Wiebe, and Rowe (2005), from which it follows that genes may be responsible for approximately 65% variability of exposure of the individual to friends and peers, who are smoking and drinking, while the influence of the environment is much lower. Another study shows that genetic factors are responsible for about 20% of the variability of individual life events (Kendler, Neale, Kessler, Heath, & Eaves, 1993). It gives objective evidence that some stressful life events that happen to people "by chance" are only seemingly random, and that their own genetic predispositions may have a subtle, but still a detectable influence on the probability of some adversities or disasters.

Such studies must be taken with caution so far as they are based on statistical associations, while each case is unique. Nevertheless, all correlations which are based on human behavior and involve stress-reactivity may, of course, have implications for the formation of psychopathologies and suicide. Moreover, in the case of suicidal behavior, interactions and covariations that may occur may be even more complicated. In one of his works, Kendler (2010) proposes ten interactive pathways that may be relevant in suicide: (1) direct effects from psychiatric disorders; (2) direct effects from personality; (3) direct effects of early adversity; (4) direct effects of current adversity; (5) indirect effects of genes on selection into adversity (gene–environment correlation); (6) interactions between genetic risk and current adversity as an example of gene–environment interaction; (7) interactions between early and current adversity which may be referred as environment–environment interaction; (8) interactions between culture and genes; (9) dynamic developmental pathways involving causal loops from genes to environment and back again; and (10) gene to environment and to development interactions (Kendler, 2010). The last type of interactions together with the idea of causal two-sided loops that link genes, environment, behaviors, and development is of utmost importance so far as it prepares a background for our modeling.

Most of the behavioral and psychological phenotypes are polygenic, that is, determined by big sets of genes, some of which are pushing toward one extreme end of the trait or behavioral pattern, while other may be pushing in an opposite direction. The existence of GxE (in contrast to rGE) is associated with the fact that in the majority of cases it is not the trait or behavior, which is transmitted genetically, but rather "norm of reaction," that is, the sensitivity of the biological system to environmental cues. Extremely strong and negative (threatening) environmental signals have a higher potential for engaging genetic susceptibility, though normal or neutral signals are less potent, but may also have significance. Sometimes the presence of a certain allele (or constellation of alleles) means enhanced or inadequate response to environmental social cues, which may be referred as quite normal. For instance, in different social situations, almost all people consume alcohol, but alcohol problem develops only in the certain percentage of alcohol consumers. The strength and the modality of the environmental signal (favorable or unfavorable, stressful or supportive, combination and balance of stress and support, etc.), as well as the period during which the environment acts (gametogenesis or fetal development, early postnatal period, early or late childhood, puberty, maturity, aging) is of great importance. All psychological and behavioral functions, as well as psychopathologies in humans, are emerging during development, in the process of nurturing, growing up, and socialization. In general, the earlier is the negative impact, the worse are the consequences because the brain is still developing. Recent studies on stress epigenetic programming of behavioral disturbances and psychopathologies have shown that adversities often induce disorder only if the influence occurs during rather short-time window. Thus, a three-dimensional model GxExD or GxExT (where D stand for development, and T for timing) can be proposed that includes genetic susceptibility or specific response rate due to the genetic make-up of the organism, the influences of the environment (negative or positive), and time of exposure (Bale, 2014). Due to objective reasons, negative effects are studied better, while the investigation of the factors of resilience, based on positive influences, is just beginning.

Evaluation of GxE is associated with modern possibilities of genetic polymorphisms assessment. In these studies two main approaches are utilized: candidate genes approach and genome-wide scanning. Candidate gene approach is based on earlier biochemical and physiological investigations, which were aimed to identify biological markers of suicidal behaviors. It is targeted on genes, which are reasonably suspected of being important in suicidality. In contrast, genome-wide studies are more like an attempt to reveal interesting and promising associations by wide coverage of the genome in relevant cohorts or populations. Earlier studies have identified several neurobiological mechanisms that may be involved in suicidal or self-aggressive behavior. They include abnormalities in serotonergic and monoaminergic systems of the brain, peculiarities of GABA-system, system of neurotrophins, and other factors of brain cellular integrity control, as well as the neuroendocrine system of stress response, immune system, and some general biochemical mechanisms associated with lipid metabolism (Mann, 1998; Rozanov, Mokhovikov, & Wasserman, 1999; Joiner et al., 2005; Wasserman, Sokolowski, Wasserman, & Rujescu, 2009; Sarchiapone & Iosue, 2015).

Primarily much attention was focused on neurotransmitter systems of the brain, so far as existing knowledge on their role in behavior, emotions, and cognition could provide a logical explanation of their possible role in suicide. Thus, serotonin system of the brain plays an important role in behaviors, emotional reactions, and cognition. It is involved in impulsive-aggressive reactions, appetite, sexual behavior and takes part in the regulation of biological rhythms and sleep, as well as in HPA regulation. It is also involved in cognitive functioning, including memory and learning. Serotonin is widely represented in the body and brain, brain stem, and prefrontal cortex that is, abundant of serotoninergic neurons. In humans, the high level of serotonin and active serotoninergic mediation in the brain are thought to be associated with optimism, confidence, and relaxed mood. People with high serotonin turnover are controlling their impulses well, are farsighted, self-critical, and conscientious. On the contrary, low serotonin tone means aggressiveness, competitiveness, and impulsive reactions. Low serotonin is also associated with fearfulness and inhibited emotional state (Wasserman, 2006).

Two other important systems are dopamine and noradrenaline systems. Dopamine neurons in basal ganglia are involved in motor activity, while in the limbic system they regulate reward-motivated behaviors. Dopamine is also involved in cognitive processes, especially attention and motivation. High dopamine level means hyperactivity and impulsivity, sensation seeking and appetent behavior (interest and excitement), while lowered or compromised dopaminergic regulation is the main reason of anhedonia (Wasserman, 2006). Noradrenaline system besides its active participation in stress reaction takes part in cognitive processes by analyzing information and screening of important or superfluous signals. High noradrenaline levels are associated in humans with dependence on reward and personal well-being and hypersensitivity to criticism. Individual who are prone to depression have usually low activity of noradrenaline system (Wasserman, 2006). GABA-system is the main inhibitory neurotransmitter system of the brain and, as recent data testify, may be involved in depression, while lipid metabolism disturbances (lowered cholesterol and fatty acids) seem to provide nonspecific background for neurotransmitter systems disturbances by impaired state of biological membranes or through proneness to risky behaviors due to reactivity to lowered blood level of these important nutritional molecules (Rozanov et al., 1999).

No surprise that intensive search of genetic factors associated with suicidal behaviors and G×E interactions was focused on previously mentioned systems, serotoninergic mediation being the most attractive subject, given its role in depression. Early studies searching for neurobiological correlates of suicidal behavior have revealed that in various regions of the brains of suicide victims the level of serotonin (5-hydroxytryptamin, 5-HT) is critically lowered (Shaw, Camps, & Eccleston, 1967). Later signs of lowered metabolic turnover of 5-HT were revealed in the cerebrospinal fluid of depressed patients with violent suicide attempts (Asberg, 1997). These data have been confirmed in many clinical-biochemical and genetic studies, firmly establishing that serotonin system of the brain is deeply involved in the pathogenesis of depression (Wasserman et al., 2009). Quite recently extensive analysis of literature has confirmed a clear association between low cerebrospinal serotonin and impulsive aggression and suicide (Glick, 2015). In genetic studies of depression and suicide alleles associated with all components of 5-HT mediation—tryptophan hydroxylase as the main regulated enzyme of serotonin biosynthesis, subtypes of serotonin receptors, serotonin transporter responsible for reuptake of the neurotransmitter, and major enzymes of serotonin degradation (catecholamine-O-methyltrasferase and monoamine oxidase, COMT and MAO) were studied and tested (Mann et al., 1997; Arango, Huang, Underwood, & Mann, 2002; Zalzman, Frisch, Apter, & Weizman, 2002). At current stage these studies have accumulated great factual material on associations of different alleles of 5-HT components with different signs of suicidality, from suicidal ideation to completed suicide (Wasserman, Sokolowski, Wasserman, & Rujescu, 2009; Sarchiapone & Iosue, 2015). Moreover, studies continue to reveal new biological system of the brain which appeared to be linked to suicidality, for example, neurotrophins (BDNF), angiotensin-converting enzyme system, GABA receptors, estrogen receptors, cholecystokinin, substance P, and other factors (Bondy, Buettner, & Zill, 2006; Rujesku, Thalmeier, Moller, Bronisch, & Giegling, 2007; Wasserman et al., 2009). Some revealed associations at first glance seem unrelated to suicidal behavior. For example, an association of suicide attempt with polymorphisms of genes responsible for some basic nerve system functions, in particular, with one of the subunits of Na^+ channel (SCN8A) and protein, associated with synaptic vesicles and involved in the synaptic cycle (VAMP4) was found (Wasserman, Geijer, Rozanov, & Wasserman, 2005). Such findings may depict the fact that relations between such complex and uncertain phenotype as suicidality and genetic factors are mediated by multiple and extremely complex intermediary endophenotypes—biological mechanisms that serve the basis both of personality traits (impulsiveness, aggressiveness, excitement or novelty seeking, negative emotions, etc.), psychopathologies (depression, adjustment disorder, personality disorder) and more subtle factors, such as cognitive style (rigid or tunnel thinking, impaired decision making under stress, etc.) (Fig. 5.3). Endophenotypes may have closer relationships with genetic factors that ultimately give rise to psychopathologies, traits, or behaviors, therefore finding the genetic basis of such endophenotypes partly may be easier than establishing links between genes and behaviors. On the other hand, endophenotypes may also have a very complex genetic background (Flint & Munafo, 2007).

Since it has been well established that depression is associated with disturbances in the serotonergic system of the brain in the search of G×E much attention has attracted gene for serotonin transporter (5-HTT). The reason is the so-called 5-HTT linked promoter region (5-HTTLPR)—a polymorphism revealed in this gene, which has several remarkable features. It is located in the promoter region and is functional, that is, has direct effect on the amount of the protein synthesized (it must be noted that in many cases genetic variations touch very local sites that may be not directly related to specific gene, only neighboring to it, and therefore their functional role is often not clear). 5-HTTLPR is associated with a repetitive sequence which may have insertion or deletion of 44 base pairs, resulting in short (S) and long (L) alleles. It is known that S-allele is responsible for reduced 5-HTT transcription and, consequently, reduced serotonin reuptake and serotonin responsivity in the synapses, ultimately leading to the reduced level of serotonin metabolite 5-HIAA in the nerve tissue and cerebrospinal fluid (Wasserman et al., 2009). Thus, these two existing variants

Phenotypes
• Behaviors, traits, or disorders
• Cognitive peculiarities, perceptions, appraisal, and reappraisal

Endophenotypes
• Thousands of biological mechanisms
• Hundreds of developmental pathways

Genotype and epigenotype
• About 20,000 genes
• Myriads of changing epigenetic marks in different tissues

FIGURE 5.3 Genome and epigenome are linked to behaviors and mental health through complex biological mechanisms known as endophenotypes.

of 5-HTT gene may result in a differential tone of the serotonin system in the population, represented by three genotypes presumably leading to low (SS), medium (SL), and high (LL) activity of the transporter. The interest toward serotonin transporter protein is based also on its practical importance—it is the main target for the most widely spread medication for depression, the selective serotonin reuptake inhibitors (SSRI). As a result, genetic variations of 5-HTT is currently one of the most well-studied genetic mechanism in molecular psychiatry and suicidology.

This polymorphism, as well as serotonin binding to 5-HTT, was studied in suicidological context by many authors and in different study designs. In postmortem studies, there was found lower binding of the ligand to 5-HTT in the prefrontal cortex of suicide victims (Mann et al., 2000). In suicide attempters, the role of reduced 5-HTT activity and association of short allele with violent suicide attempts was also confirmed (Lin & Tsai, 2004; Wasserman et al., 2007a). These and many other studies have established that 5-HTT may be a critical point in the whole system of serotonin innervation in brain that it is associated with depression and suicide and that state of serotonin system may be referred as genetic-based trait, which is important for development of suicidality, while other systems are more context-dependent (Rozanov, 2013).

The most famous study focused on this polymorphism is the longitudinal study dedicated to the role of stress in depression and suicidal activity conducted by Caspi et al. (2003). In this study, 847 subjects have undergone 9 psychosocial surveys between the ages from 3 to 26 years. Based on genotyping data participants were classified into groups in relation to 5-HTTLPR polymorphism. Among the participants, 147 appeared to be homozygous carriers of the (S) allele, 265—of long (L) allele, and 435 were heterozygous (SL) in full accordance with typical population genetics distribution. The frequency of 14 types of negative life events in the age from 21 to 26 years (finances, housing, unemployment, health, relationships, etc.) were assessed together with symptoms of depression, the presence of the diagnosis of depression in the last year, the fact of suicide attempt and severity of suicidal thoughts. It was found that homozygous carriers of the short allele are most likely to develop subclinical and clinical depression in response to accumulation of life stress which is consistent with neurobiological mechanisms.

But the carrier state of specific alleles appeared only part of the vulnerability. The most impressive finding of the study was revealed when regression analysis estimating the association between childhood maltreatment (between the ages of 3 and 11 years) and adult depression (ages 18–26), as a function of 5-HTT genotype was performed. It was found that probability of depressive episode is substantially higher in those subjects who have experienced severe damaging stress in early childhood. Thus, the short allele, interacting with the environment in a comparatively short time window (early childhood) affects the risk of depression and suicide in further adult life (when negative life events accumulation becomes more probable) in genetically vulnerable individuals. Thus, childhood stress predicts adult depression mostly among individuals carrying at least one S allele (Caspi et al., 2003).

These results have promoted many other studies of GxE in relation to suicide focused on the same and other relevant brain systems, which confirmed these general regularities and relationships. Thus, Roy, Hu, Janal, and Goldman (2007) demonstrated the interaction of childhood trauma with 5-HTTLPR low expressing allele in patients with substance abuse predicting higher risk of suicidality. Brezo and colleagues in a longitudinal (22 years) study have evaluated probabilities of suicide attempts associated with 143 single nucleotide polymorphisms in 11 serotonergic genes, acting directly or as moderators in gene–environment interactions with childhood sexual or physical abuse. It was found that HTR2 (5-hydroxy-tryptamine receptor 2A) gene variants were involved in suicide attempts by interacting with the history of child abuse (Brezo et al., 2010). Perroud and coauthors have found an interaction between childhood trauma and well-studied BDNF gene polymorphism (Val/Val variant) in a prediction of violent suicide attempts (Perroud et al., 2008). In several other studies, interactions between early life adversities and genes involved in HPA regulation were found. Wasserman, Sokolowski, Rozanov, & Wasserman, 2008 have described GxE for CRHR1 (CRH receptor 1) gene which was sex-specific—association with suicide attempts and interaction with adolescent stress was found in females and in males with low stress for different SNPs. Roy, Gorodetsky, Yuan, Goldman, and Enoch (2010) were looking for an interaction of FKBP5 gene with childhood trauma and attempted suicide (Roy et al., 2010). This gene product acts as the modulator of CR binding to its receptor and its variants are known to interact with childhood abuse in the prediction of PTSD (Binder et al., 2008).

Thus, while classical behavioral genetics studies have proved that there is a detectable genetic input in different forms of suicidal behavior, candidate genes approach have identified several important loci that are strongly associated with suicidal behavior. Most of the genes that appeared important were actually suggested by previous studies that were able to elucidate the role of brain neurotransmitters and extraneural systems in suicidal behavior. There were big hopes regarding genome-wide studies and they have really identified several interesting loci but how they may be related to suicidal behavior remains mostly unclear (Wasserman et al., 2009; Sarchiapone & Iosue, 2015). These findings may influence future development in the neurobiology of suicide. To our mind biggest success in understanding suicide was achieved in the way of evaluation of GxE interaction, as well as GxExT and GxExD modeling. It has been well established that early life adversities have negative consequences regarding mental health problems later in life, including self-harm and suicide. Now it looks very probable that early life stress (or psychological trauma as it is usually understood by psychologists and psychiatrists) substantially enhances suicide risk associated with genes related to critical brain neurotransmitters and other key brain systems and neuroendocrine mechanisms. Therefore suicide risk can be understood as the genetic predisposition (diathesis) based on genetic make-up, specifically on having constellations of alleles that determine impaired functioning of critical brain systems (mainly serotoninergic) and enhanced reactivity to life stress as a consequence of serious adverse influences in childhood (Fig. 5.4).

The combination of negative environmental signals and existing predispositions create persistent biological and behavioral response system, which can be referred as a stress-vulnerable endophenotype. Development of such phenotype means enhanced predisposition to mental health disturbances and suicidal behaviors throughout life. This endophenotype may comprise different biological mechanisms starting from hypersensitivity of HPA, SAS, and continue with impaired neurogenesis and cellular disturbances in the hippocampus and prefrontal cortex and enlargement of the amygdala. All these features may provoke a variety of emotional and behavioral disturbances including propensity to depression and anxiety. It also may influence cognitive peculiarities and weaken control over aggressive (or self-aggressive) impulses. All this actually contributes to stress-vulnerability, which manifests itself when a usual psychosocial stress of life (negative life events, break of relations, crises, frustrations, loneliness, perceived burdensomeness and other triggering events) start their action. Stress experienced by the personality due to cognitive reappraisal in interaction with existing vulnerability leads to further deteriorating effect of stress hormones which ultimately results in deeper disturbances of neuronal structures and brain circuits and behavioral and emotional disorders as well as sleep problems. Problems of cognitive style, such as rumination, thoughts agitation, problems solving impairment, tunnel thinking, and feeling of being trapped (which may also be the consequence of early trauma) may add to existing risks and may make suicide more probable. The model presented at Fig. 5.4 also implies that childhood passed under positive conditions can provide a positive mental health, even in the presence of the genes of vulnerability. Moreover, the same genes in diametrically different environmental conditions can cause vulnerability and be protective. This mechanism emphasizes the interaction of genes and environment—a key interaction which may be important for the probability of suicide in the lifetime.

In another word, negative early adversities lead to the formation of "incubated trauma," the consequences of which can be seen later in life. One of the most intriguing questions is how this early psychological trauma is "conserved" and kept more or less silent until first usual life difficulties

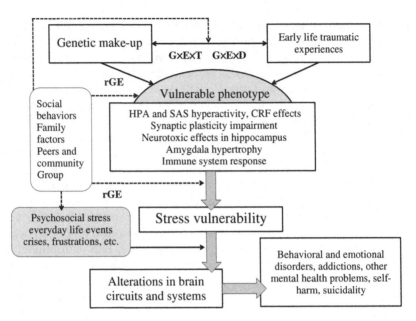

FIGURE 5.4   Genes to environment to time and development interactions, behavioral genes to environments covariations, and stress-vulnerability.

emerge in puberty and early adulthood. There is a growing body of evidence that epigenetics—the way how "environment talks to genes" is just one of the important mechanisms. It is, therefore, important to have confirmation of this idea directly from subjects with suicidal behavior, including completed and attempted suicide. It is important to know if such epigenetic events really take place, what genes or genetic loci are involved, what cells and tissues are impacted, can these marks be found only in brain or in other tissues, etc. Epigenetics is very seducing—it has the tendency to explain everything, therefore every hypothesis or suggestion about the participation of epigenetic mechanisms in the propensity to any behavioral pattern and suicide, in particular, should be supported by objective knowledge.

EPIGENETIC FINDINGS IN SUICIDE

Epigenetics of suicidal behavior is a growing arena of research though there are several limitations and problematic issues. Studies of this kind are using postmortem brain samples of people who committed suicide, which is associated with ethical aspects and methodological challenges (the shortest possible and standardized time from the moment of death to the moment of extraction of the material, pH control, a small number of samples, etc.). Currently, in this field, we are observing the process of accumulation of information. For example, a systematic review of epigenetic findings in suicide published in 2012 has identified 12 papers devoted to this issue (El-Sayed, Haloossim, Galea, & Koenen, 2012). Since then the number of studies has increased but this has not yet led to a breakthrough. Existing studies are related to several basic directions—epigenetic markers of BDNF and its receptor genes, studies of polyamine metabolism, studies of several glial cell marker genes, neurotrophins, serotonin system genes, GABA system genes, and HPA related genes, mainly cortisol receptor gene.

The most well-known and widely cited series of studies was aimed to prove the epigenetic nature of the incubated trauma and its role in suicide. In these studies, DNA methylation pattern in the samples of hippocampi of suicide victims was compared with those who died from other causes (control group). Victims of suicide were grouped on the basis of the fact of severe stress experienced in childhood. In order to verify early trauma, the method of the psychological autopsy was used (questioning of relatives and other informants) and these data were confirmed by analysis of medical and other available documentation. In one of the studies if was found that hippocampal rRNA (ribosomal RNA) was significantly hypermethylated throughout the promoter and 5′ regulatory region in the brain of suicide subjects who have suffered severe adversities in childhood. It was accompanied with reduced rRNA expression in the hippocampus, suggesting that proteins synthesis machinery may be impaired in neurons, consistent with known findings of reduced volume of the hippocampus in abused individuals (McGowan et al., 2008). These differences in rRNA methylation were found particularly in the hippocampus, but were not evident in other parts of the brain, for instance, in the cerebellum. It also occurred in the absence of genome-wide changes in methylation, suggesting the specific location of epigenetic marks. In another study, it was found that the expression of the GR in the hippocampal tissue of suicide victims with the history of childhood trauma was specifically reduced in comparison with both those who committed suicide and were not traumatized and those who died of other causes (McGowan et al., 2009). These data provide evidence that early childhood events in humans alter the epigenetic landscape of the brain, actually being the first known confirmation of animal (rodents) studies of early life stress effects. It also supports the idea that epigenetic processes mediate effects of social adversity during childhood on the brain and produce alterations that persist into adulthood and enhance suicide risk (Labonte & Turecki, 2010). Both previously mentioned studies are the most cited in the field of suicide epigenetics.

Several studies on epigenetics of suicide are dedicated to neurotransmitter systems of the brain, particularly 5-HT and GABA system. For instance, in the frontal lobe tissue of postmortem brains of suicide victims with psychosis hypermethylation of promoter DNA of one of the types of serotonin receptors HTR2A was found which coincided with dysregulation of this receptor and other candidate genes on the 5-HT system (Abdolmaleky et al., 2011). The same pattern of hypermethylation of this receptor was found in peripheral leucocytes of suicide attempters with schizophrenia (DeLuca, Viggiano, Dhoot, Kennedy, & Wong, 2009). Studies of GABA system up to now are limited with findings of downregulation of DNA-methyltransferases mRNA expression in frontopolar cortex suicide victims, which coincided with hypermethylation of GABAA receptor and reduced expression of this gene (Poulter et al., 2008). These studies give the impression that main brain systems' components are touched by epigenetic transformations in suicide, implicating that these changes may play the role in the development of suicidal behavior (Bani-Fatemi, Howe, & De Luca, 2015).

Some studies are utilizing wider approach when DNA methylation is studied on a genome-wide scale. For example, methylation of CpG sites in the ventral prefrontal cortex of 25 suicide victims with a confirmed history of severe depression compared with 28 controls who had died from other causes and who had no diagnosis of depression was studied. Age of subjects was within the range from 16 to 89 years. It has been shown that methylation in the brain increases in older subjects (this is contrary to reported reduced methylation with age), while in the brains of persons who committed suicide the degree of methylation was approximately 8 times higher (Haghighi et al., 2014) The authors conclude that this wide methylation contributes to the increase of suicide risk in affective disorders, especially in the process of cellular aging, which is associated with oxidative stress and neuron-glial relations disturbances. In another methodologically similar study, enhanced DNA methylation in the prefrontal cortex of suicide victims was revealed in a large number of genes. Though global methylation of all CpG sites appeared to be lower than in control brains, top differentially methylated regions were higher in suicide victims. Authors consider that differential methylation and epigenetics, in general, is involved in the pathophysiology of suicide given the role of the frontal lobes in the decision-making and future planning. Moreover, identified top methylated genes pointed to a number of existing and new candidate genes that may be involved in suicidal behavior, including BDNF gene (Schneider, El Hajj, Müller, Navarro, & Haaf, 2015).

So far as there is a close relationship between depression and suicide, epigenetic modifications in depression and suicidal behavior have been reviewed in detail from the point of view of a possible role in both (Lockwood, Su, & Youssef, 2015). Authors pay special attention to hypermethylation of the BDNF gene and its receptor TrkB (tropomyosin-related kinase B) and other neurotrophins and conclude that epigenetic events serve as a platform for interaction between genes and stressful environments in the formation of affective disorders and suicidality (Lockwood et al., 2015). What attracted much attention was a decrease of BDNF mRNA expression level in brains of suicide victims, which was most likely due to hypermethylation of promoter region of exon IV of the corresponding gene in Wernicke's area of the brain (Keller et al., 2012). Hypermethylation was also found in several sites of the promotor of TrkB gene which was associated with lower TrkB expression (Ernst et al., 2009).

Several other studies are discussing the genome-wide pattern of epigenetic markers in glial cells in depression and suicide (Nagy et al., 2015) and the epigenetic nature of polyamine metabolism disturbances in the brain in depression and suicide risk (Gross, Fiori, Labonté, Lopez, & Turecki, 2013). Interesting data were obtained regarding the epigenetic regulation of the metabolism of polyunsaturated fatty acids in suicide attempts on the background of depression. Many studies provided data that dysregulation of metabolism of fatty acids and cholesterol, or rather deficiency of them in the body, is associated with neuropsychiatric disorders and suicidal tendencies (Rozanov & Mid'ko, 2006), while recent data suggest a link between hypermethylation of the gene responsible for synthesis of elongase, an enzyme involved in elongating fatty acid chains with suicide attempts (Haghighi et al., 2014).

Within the model that supports leading role of glucocorticoid receptor and its modulators in stress, depression, and suicide data were presented recently on epigenetic modification in nerve cells of the prefrontal cortex of spindle and kinetochore-associated protein (SKA2) gene, involved in the regulation of cortisol binding to its receptor and its interaction with the nuclear structures. This protein (chaperone), which is also involved in the process of chromosome segregation during mitosis, is reduced in the prefrontal cortex of suicide victims resulting in impaired control of HPA activity and cortisol levels under stress. Its methylation level predicts suicidal behavior and PTSD when assessed not only in the brain but also in the blood (Kaminsky et al., 2015). This was the basis for the authors to declare test for suicidality, predicting the risk with the 90% probability (Le-Nicilesku et al., 2013). SKA2 appears to be a very promising gene involved in suicide and PTSD; recently the same group has published results of a translational study that links together this gene, mood, stress, and longevity, quite possibly identifying the molecular convergent point of mind–body interactions (Rangaraju et al., 2016).

This short review gives an impression that epigenetic transformations, mostly changed DNA methylation patterns are systematically identified in connection with suicidality. Quite recently several reports have been published where this line of research is called one of the most promising both for further study of the biological mechanisms of suicidality and development of tests to evaluate the increased risk as well as for finding promising targets for correction and prevention strategies. Epigenetics is seen by many authors as a link between environment and genome in relation to mental health, behavioral disturbances, and ultimately suicide (Nagy & Turecki, 2012; Turecki, 2014; Bani-Fatemi et al., 2015; Rozanov, 2015).

Epigenetic findings in suicide are suggesting several important conclusions. It becomes clear that epigenetic transformations and related changed expression of certain genes are common events in the nervous tissue of suicide victims. Localization of epigenetic marks (hippocampus, prefrontal cortex) is consistent with the role of these structures in emotional and cognitive impairments that are inherent to suicidal individuals. On the other hand, these structures are given priority while searching epigenetic marks in suicide victim's brains, and it does not mean that they can be found in a much wider set of structures. Several studies provide very important evidence that epigenetic marks are associated with early life adversities and were persisting in brains until suicidal act. Epigenetic transformations are found in promoters of genes that belong to several brain systems that are involved in suicide pathogenesis (serotonin and GABA neurotransmitter systems). They are found within HPA genes, in particular, GR gene and within BDNF and its receptor genes. It confirms that stress plays an important role in the suicidal act and that it may be associated with a variety of changes in cellular dynamics in the developing brain. On the other hand, changed methylation pattern is often found in different genome loci when studies are performed in a system-wide manner. Most frequently noticed methylated genes in such studies often seem rather far from known pathogenetic pathways of suicide suggesting the existence of biological mechanisms associated with suicidality that until recent times remained obscure. It looks that revealing of postmortem epigenetic marks may be more useful in terms of further understanding of the pathophysiology of suicide than candidate genes and genome-wide studies. Some epigenetic marks can be identified in peripheral blood cells which raise hopes of biological predictive test for suicidality. Such test is already reported, but the results need replication. Epigenetics is an important link between changing environment and conservative genome. It suggests a mechanism by which changing environment can produce very fast biological, structural, and behavioral changes, which may have implications for observed changes in prevalence of mental health problems and suicide in young people.

BIOBEHAVIORAL-PSYCHOLOGICAL-EXISTENTIAL MODEL OF SUICIDE

We have set a goal to integrate social, psychological, and biological aspects of suicidal behavior in a comprehensive model which will be able to suggest an explanation of the growth of suicides in the young people observed recently. Epigenetics, with regards to all accumulated knowledge, which is overviewed in previous chapters, is an excellent candidate for the central mechanism that unites these domains. Any integrative model of suicide is always a generalization or speculation and there is a great probability that the result will be either incomplete or skewed. We are aware that it is a complex task, and there is also certain skepticism that such model can be developed in general, given that people who commit suicide constitute a very heterogeneous group. Thus, we should keep in mind variability of reasons, pathways, and mechanisms that eventually may result in a suicidal act. On the other hand, it is quite possible that it is epigenetics with its potential for embedding short-term environmental cues and reproducing acquired traits or behaviors in generations that may contribute to enhancing this heterogeneity. Thus, through different pathways and within different mechanisms, but probably always involving epigenetics, recent generations of young people are becoming more suicide prone, which results in growing of a general number of suicides in this age group.

Our attempt to develop an integrated model is based on several productive concepts and ideas. One of this concepts, which is strongly supported by factual data on the role of epigenetics in early life programming, periodization of development, and predictive role of early transition periods, is the concept of Developmental Origin of Health and Disease, DOHaD (Hochberg et al., 2011; Kubota, Miyake, Hariya, & Mochizuki, 2015). This concept gives a frame for the whole discussion because it puts forward the role of early life programming as the main mechanism by which existing predispositions are enhanced due to stressful influences during sensitive periods of development. From the point of view of this concept many human age-related diseases are actually pediatric disturbances so far as susceptibility to them is built in rather early periods of life. This is especially true for mental disorders and most prevalent mental health problems (anxiety, depression, addictions), which also need a certain time for their development. Epigenetic mechanisms involved in mental health programming are the most widely discussed today, especially in relation to stress (Szyf, 2011; Lester, Marsit, Conradt, Bromer, & Padbury, 2012). Within the frame of DOHaD concept, all medical knowledge on diseases prevalence and timing of their emergence are seen in a new light. Recently, there was a suggestion that this concept may be supplied by adding "behavior" in the list of programmed factors, that is, "Developmental Origin of Behavior, Health and Disease", DOBHaD (Van den Bergh, 2011). Central to this hypothesis is interdependence of developmental influences, genes, environment, and behavior, which is especially important with regards to choosing healthy or unhealthy behaviors by young people at turning points in their life. Human behavior is the result of complex interplay between biological, personality, and social factors, and very often choosing unhealthy behavior is the result of chronic psychosocial stress, when loss of hope and pessimism regarding future is unconsciously converted into behaviors that may give an immediate reward but which in longer run actually mean life-shortening strategies. Such strategies were described by N. Farberow as indirect self-destruction or continuous suicide (manifesting itself in drinking, heavy smoking, risk-taking, excitement seeking, etc.) which may be exacerbated by the acute suicidal act at some moment (Farberow, 1980). All these types of behaviors are often referred as risky behaviors, which are often associated with suicide, especially in youngsters (Wasserman et al., 2010). From this point of view DOBHaD concept is quite relevant for understanding suicidal tendencies. It also adds to our understanding of the evolution of human behavior in response to changing environment in comparatively short historical periods of time. The focus on development and evolution is very important so far as it evaluates outcomes of behavioral strategies in a wider context. Interaction of evolution and development (Evo-Devo perspective) provides a most plausible explanation how some negative temperamental traits and behaviors may appear adaptive in quickly changing environments under growing stress pressure, while being maladaptive in traditional and more secure environments, and vice-a-versa (Lester, Marsit, Conradt, Bromer, & Padbury, 2012).

We are also using "suicidal process" and stress-vulnerability model of D. Wasserman as basic biopsychosocial concept mostly relevant for the integrated approach we are promoting. From this model, we borrow understanding of the suicidal process as continuing series of events, during which risk and protective factors are constantly counterbalancing each other until some unfortunate trigger change the balance in favor of self-destruction. At last, we keep in mind environment-based approach which incorporates epigenetic mechanisms as an immediate response to changing environment and which encounters the possibility of transgenerational transmission of epigenetically acquired phenotypes (Golubovskii, 2000; Nazarov, 2007; Keverne & Curley, 2008). In general acknowledging of dominating role of environment in "nature or nurture" or "genes or environment" dichotomy and the pivotal role of different types of interactions (genes, environment, development, and timing) are the basics of our modeling. Rather short time during which identifiable changes in suicide indexes occur (several decades) is best explained by epigenetic, not genetic changes in the population, though genetics also plays its role.

Several years ago we have formulated a position that early periods of life development to a certain extend predict suicidal behavior of the individual, thus pointing on neurodevelopment roots of suicide (Rozanov, 2010). We have stressed, that one should take into consideration several levels of determination of suicidal behavior, which are interacting with each other. The first level is based on an unfavorable combination of genes, predisposing to certain temperament, and personality traits (neuroticism, angry and hostile reactions, depression, impulsivity, stress-vulnerability, excitement seeking, etc.). On this level, dominating interaction is G×E and existing examples with 5-HTT or BDNF confirm it. On the second level, main role belongs to early stages of development. In a case of early life adversities, child neglect or abuse, stress-dependent epigenetic events may lead to the formation of "vulnerable phenotype" or "incubated trauma." In this case, G×E×D or G×E×T interactions dominate, making the picture more complex. The third level involves behavioral pattern, due to which individual with enhanced stress reactivity may appear in recurring life situations of stress (active genes and environmental covariance, rGE), thereby setting a stereotypic response, which in a long run may make life unbearable and enhance ability to overcome fear of pain and possible death, according to Joiner's theory. One should also take into consideration cognitive component, problems solving practices, decision-making processes, the role of emotions and their interactions with cognitions, coping skills, autobiographic memory functioning, as well as even more deep and subtle factors, such as values, life purposes, meanings, beliefs, cultural norms, and traditions. Many of these factors are also partly dependent on biological background of the individual and personality, but to a greater extent are acquired during personality development and socialization. On the other hand, they may be decisive in some part of suicides. All these levels of determination, as well as interactions between them, should be evaluated in the context of the growing pressure of psychosocial stress that is natural for modernization and is associated with social inequalities and lifestyles based on competitiveness, individualism, consumerism, and hedonism. Another important factor that we will discuss within this context is perceived stress, which is based on a cognitive appraisal of stressful situations, subjectively perceived social conditions, and imaginary threats evaluations.

Since a lot of objective knowledge has been accumulated in the domains of stress neuroendocrinology and behavioral and psychiatric epigenetics, the topic of the neurodevelopmental origin of suicide has received a new impulse. Quite recently several publications overviewing state-of-the-art in this field, elaborating new hypotheses and formulating unsolved questions have been published (Turecki, Ernst, Jollant, Labonte, & Mechawar, 2012; Rozanov, 2013, 2015; Nagy & Turecki, 2012; Bani-Fatemi et al., 2015). For instance, Turecki et al. (2012) have focused on childhood maltreatment, neuroendocrinology of severe stress and findings of epigenetic modification of GR gene in the hippocampus of suicide victims' brains. Discussing distal and proximal risk factors of suicide and integrating existing knowledge from animal and human studies of epigenetic effects of environmental variations on brain and behavior the authors developed a comprehensive model explaining the higher risk of suicide in individuals subjected to early life adversities. In this scheme, early life stress through increased methylation of GR promoter leads to HPA dysregulation, which becomes the reason of a variety of physiological abnormalities both in the central and peripheral parts of the stress system. The last serve the basis for the development of impaired emotional/behavioral and cognitive phenotypes, which enhance the risk of suicidal act. Emotional abnormalities include enhancement of anxiety and depressive symptoms, while behavioral—impulsivity and aggression. Cognitive alterations include impaired decision making, poor problem solving, high emotional sensitivity, and impaired attention and reduced verbal abilities (Turecki et al., 2012).

Turecky and coauthors have pointed on the percentage of childhood adversity sufferers among individuals who display suicidal behavior. This part is ranging from 10% to 40% depending on the type of abuse and suicidal activity. In relation to this significant, though minor subgroup, Turecky and coauthors propose a pathway focused on the epigenetic programming of HPA hyperactivity from one side, and BDNF system from another side. The first system, which is more involved in behavioral and emotional reactivity to stressors of everyday life, is proposed to be responsible for anxiety, oppositional behaviors, hyperactivity, and impulsivity. The second (in functional conjunction with the first) is proposed to be responsible for cognitive deficits and abnormalities, largely due to impaired development of critical brain structures—gray matter of prefrontal cortex, anterior cingulate, superior temporal gyrus, amygdala, hippocampus and corpus callosum (Turecki et al., 2012). Thus, impaired cognitive functions are seen largely as consequences of structural abnormalities of the brain, which are caused by severe stress in the early childhood, epigenetics being the central mechanism that programs such development by changing methylation of critical genes responsible for cellular networks organization and dynamics. This model, which may be characterized as "cell-based" or "structural" model is supported by other suicidologists (Haghighi et al., 2014).

On the other hand, there are many questions that remain unsolved within this model. "Structural" model that puts forward cellular disturbances acquired during development and maturation does not explain suicides in young people who were not subjected to early traumatic events. On the other hand they are the majority; moreover, there are reports that suicides in youngsters may be associated both with high and lowered intellectual and cognitive development (Pfeffer, 2000). Of course, general intellectual and adaptive success in a modern world does not mean coping skills and good problem solving, however, if we fully accept the position that all problems come from biological mechanisms and cellular misbalances in the brain that are induced by early life stress, we cannot explain great portion of suicides and especially global growing trend of suicides in modern youngsters. Suicides in adolescents not necessary occur on the background of mental health problems; on the contrary, some suicides are associated with perfectionism, high expectations and impressive achievements, and narcissistic tendencies (Apter & Gvion, 2016; Bodner, Ben-Artzi, & Kaplan, 2006). After reviewing all existing studies on developmental influences on adolescent suicides the authors of the paper (Manteaux, Jacques, & Zdanovich, 2015) also acknowledge the importance of biological factors that predispose youngsters to suicides but advocate that we should not restrict it to structural neurological changes without taking into account the other factors of suicide. We are supporting this point of view too.

Our approach is based on the wide understanding of the role of stress and epigenetics in suicide. We would like to draw attention to the fact that across the lifespan epigenetic events may be evoked by stress of different nature, this stress may be different in utero, in early life, and in further life. The main driving force of epigenetic phenomena is stress caused by threats to basic early needs—threat to life and general security, alimentary security, parental (social) support. These fears arise in early childhood, and frustration of these needs—separation from mother, poor mothering, mother's stress and depression, loss of a parent, and ultimately child physical and sexual abuse may lead to epigenetic programming of behaviors that will be beneficial in a threatening and hostile environment—impulsivity, aggression, anxiety, etc. Natural studies testify that all above-mentioned adversities are associated with suicidal behavior (Fergusson, Woodward, & Horwood, 2000). In such cases suicidal acts may be related to dysfunctions in critical brain structures in accordance with the "structural" model (Turecki et al., 2012). Growing understanding of the origin of these problems evoke positive actions that must diminish the prevalence of such situations in human communities, and many efforts are made to achieve this goal in educational and social systems. Really, when policies are introduced and consistently followed a decrease of the problem may be achieved (Finkelhor, Shattuck, Turner, & Hamby, 2014). Nevertheless, pediatric stress persists in every society, every new generation of young adults, who are becoming parents, are producing a certain percentage of dysfunctional families with high risks of family stress, violence, and child neglect and abuse, which is depicted in reports on family violence and child abuse. International studies show that a quarter of all adults report experiencing physical abuse as children, and that and 1 in 5 women and 1 in 13 men report experiencing childhood sexual abuse. In a great portion of cases child neglect and abuse manifests itself in difficulties of bonding with a newborn, not nurturing the child, lacking awareness of child development or having unrealistic expectations, misusing alcohol or drugs, including during pregnancy, being involved in criminal activity and experiencing financial difficulties. In a great portion of such cases, parents were maltreated themselves when being a child (WHO, 2014a). It gives an impression of the prevalence of situations in which parents may provoke early childhood trauma in their children and which may have profound consequences for the mental health of their offspring during the whole future life. On the other hand, growing psychosocial stress among parents, especially mothers may have additional negative impact.

More and more recent studies bring evidence that stress, depression, and anxiety in mothers may lead to negative programming of immune dysfunctions and mental health problems in children and that transgenerational transmission of these behaviors may be based on chemical marks in the epigenome. Animal studies suggest that many anxiety-related traits and behaviors may be dependent on permanent alterations in brain transcriptome, which acts as a transgenerational phenotype, leading to further anxiety-related disorders and cognitive/memory impairments in new generations (Crews, 2010). There are confirmations that children of depressed mothers have hundreds of differentially methylated genes in the immune system, while in the hippocampi of deceased older age individuals in relation to the history of depression in their mothers a great number of coinciding genes remain methylated (Nemoda et al., 2015). Maternal anxiety during pregnancy is associated with demethylation of insulin-like growth factor and this has a relation to birth weight (Mansell et al., 2016). Prenatal maternal depression leads to more active functional involvement of amygdala in their 6-months-old children which may mean problems during socializing and higher susceptibility for developing mental health problems in future life (Qiu et al., 2015). These studies are bringing more and more confirmations that mother's mental status has profound consequences in offspring via epigenetic mechanisms.

Moreover, recent animal studies suggest that paternal stress can also cause anxiety and depressive symptoms in the offspring (Short et al., 2016). Thus, "parents' sins" are transmitted to children, and quite possibly to their children too, both due to behavioral and chemical transmission. These complex genetic, environmental, and epigenetic contributions to development of behavioral and mental health problems, together with growing stressful pressure from the environment may be the reason for the general growing of suicidal behavior, which is registered worldwide. The same reasoning is utilized by many authors while discussing life trajectories for disruptive behavior or reproducing of maternal style with subsequent growth of mental health problems (Tremblay, 2010; McEwen, 2012; Curley & Champaigne, 2016). As a conclusion, there exists a portion of children and adolescents that suffered early life adversities and have mental health consequences due to epigenetic transmission of dysfunctional parental behavior, adding constantly to existing vulnerabilities in the population.

It is also interesting to mention in this context that recent studies have pointed on sex differences in the mechanism by which brain responds to stress in males and females. Recently, this issue was discussed by McEwen, Gray, and Nasca (2015). This evidence comes mostly from animal studies, in which it was shown, that female rodents do not demonstrate the same pattern of brain neural remodeling in chronic stress as do males. Male stressed rats show more pronounced impairment of hippocampal-dependent memory functions but in some models of stress demonstrate even better performance (McEwen, Gray, & Nasca 2015). These data, though not directly confirmed in humans, are interesting to compare with sex differences in suicidal behavior, which consistently evidence of higher suicidal activity in males, including young age. Actually, higher stress responsivity in young males is a known fact, though it is usually explained by hormonal factors and more active adrenal glands functions (Stephens, Mahon, McCaul, & Wand, 2016). On the other hand fMRI studies have revealed sex-specific reactivity of limbic system structures in relation to early life stress, anhedonia and alcohol abuse (Corral-Frías et al., 2015). This topic is beyond our discussion, though linking early life stress, epigenetics of brain and suicide risk to sex and gender may be an extremely interesting and promising field of study.

While discussing early life traumatic experiences in children it is necessary to say several words about the balance between trauma and social support. Garner (2013) quite reasonably points, that in the early life of a child three types of stress versus support situations may be differentiated: (1) low stress and high support (i.e., caring mother provides soothing to a child after minor trauma); (2) average to serious stress and average support (i.e., close relatives provide support to a child in case of a loss of a parent), and (3) "toxic" stress, that is, violence against child or child abuse when perpetrators are close relatives, in the worst case—parents. This so-called "toxic stress" leaves really serious consequences, both structural and behavioral. On the structural level, it may lead to heightened stress vulnerability and cognitive and emotional disturbances, while on the behavioral level it results in a wide array of attempts to blunt the stress response, a process known as "behavioral allostasis." (Garner, 2013). This is about behaviors, such as smoking, overeating, promiscuity, and substance abuse, which may decrease stress transiently and which are known as risky behaviors. Over time they become maladaptive and result in the adoption of unhealthy lifestyles, which constitute risk factor of suicidal behavior and are even viewed as the main predictor of suicide. Given that the field of social support is narrowed in modern societies due to a family crisis and demographic modernization, it may mean skewing of the situation to the stronger effects of stress. On the other hand, here we can see signs of interactions between suicide rate and culture, as formulated by Kendler, so far as in more traditional societies with higher social cohesion, where an orphan child never remains alone until he has at least neighbors and where personal life is more exposed to the community, the impact of stress may be substantially lowered.

When a child passes from childhood to adolescence, the nature of stress changes, it is not limited any more to child–parent relationships and may be determined by broader social phenomena. Young adults at schools, but mostly in colleges are subjected to social environments with inequalities, consumerism, individualism, loneliness, information overload, lack of physical activity, urbanization, and virtualization of existence, which means chronic exposure to psychosocial stress. Youngsters cannot imagine their lives without their smartphones, which seem to open the whole world for them, but today "loneliness in the crowd" of social networks has become a real problem, by the way, associated with differential immune system genes expression (Cacioppo, Fowler, & Christakis, 2009). Moreover, involvement in social media can affect the mental health of youngsters (Pantic, 2014). One can consider it not so important, but the scale of involvement of adolescents in virtual communication is too large and may have an impact on suicidal activity. Adolescents also experience a variety of emotions reacting to their parents' stress. As a result, archaic natural types of stress of early life may interact with older age psychosocial stress, which is another relevant interaction mentioned by Kendler. In confirmation, our study has shown, that stress until the age 18 strongly predicts negative life events at maturity (Rozanov, Yemyasheva, & Biron, 2011).

Adolescence is the period when more acute understanding of perceived stress comes, the feeling of being stressed and inability to cope or to handle accumulating problems, subjective expectation of trouble, anxiety, and depression, which itself also means stress. Adolescence is the period of heightened stress-responsivity and emotional reactivity from the physiological and psychological point of view (Romeo, 2010). We consider perceived stress a very important component in our model and suggest that it is perceived stress that actually links together social, biological, psychological/cognitive, and behavioral processes. Recently, a substantial body of evidence is accumulated confirming that perceived stress traditionally measured by Perceived Stress Scale produces biological effects of the same kind, as stress caused by real adverse events or traumatic experiences. For instance, in students in Egypt and UK perceived stress strongly predicted health outcomes, such as pain, gastrointestinal, circulatory, and breathing systems problems (El Ansari, Oskrochi, & Haghoo, 2014). Perseverative thoughts and worries in adolescents are associated with negative health outcomes, perceived stress being the mediator of the effect (Kokonei et al., 2015). In our study, severity of depression in students is predicted by a subjective feeling of stress better than the number and severity of accumulated life events (Rakhimkulova & Rozanov, 2015). The stress-related activity of EEG depends on the level of subjectively reported stress (Luijcks, Vossen, Hermens, van Os, & Lousberg, 2015). The authors of the last study called it "proof-of-principle study" understanding very well that they are testing the biological effect of stress/coping interaction. Thus, as we have also mentioned in Chapters 2 and 4, internal and subjective feeling of inability to cope with a problem, which is understood as stress from the cognitivist point of view, has distinct biological correlates (Luijcks et al., 2015). Perceived stress is mediating influences of personality type and cultural and psychosocial factors (ethnic discrimination, family conflict) on depressive symptoms, cigarette smoking, and nonsuicidal self-injury in adolescents (Kiekens et al., 2015; Lorenzo-Blanco & Unger, 2015). It is also interesting to mention, that perceived stress is associated with problematic Internet use, corumination via cell-phones and Facebook behaviors, showing how modern IT instruments and social media are readily used to fulfill the needs of stressful communications, addictive behaviors, and anxious social contacts (Park, 2014; Murdock, Gorman, & Robbins, 2015; Morin-Major et al., 2015). Moreover, a number of Facebook friends was positively correlated with cortisol systemic output (Morin-Major et al., 2015).

The most interesting question is whether perceived stress can induce epigenetic modifications in a similar manner as it is observed in cases of mother stress, early life stress or child abuse and neglect at a very young age. From the most general point of view, taking into consideration dynamics of stress hormones and other biological mediators of stress, it is very probable that perceived stress and stress due to an identifiable external event must have the same nature. Actually, the so-called early "toxic" life stress in the form of child abuse is associated not with the event itself, but with the fact that perpetrators are close relatives, that is, undermining the main feeling of safety (Turecki et al., 2012). It may be said that this kind of childhood stress is also an imprinted subjective stress, stress "in the head," an imprinted feeling of injustice and sorrow. Not violence itself, but more a cognitive pattern of the violence is really important.

There is already indirect and direct evidence that perceived stress may trigger epigenetic mechanisms. One comes from the study in which methylation of the BDNF gene promoter is closely associated with suicidal thoughts, and not to the number of real-life adverse events (Kang et al., 2013). Another comes from the study of Lam et al. (2012) who were measuring DNA methylation in the promoter regions of more than 14,000 human genes in peripheral blood mononuclear cells collected from a community-based cohort stratified for early-life socioeconomic status. It was found that psychosocial factors, such as perceived stress, and cortisol output were associated with DNA methylation, as was early-life socioeconomic status (Lam et al., 2012). It is remarkable that changed methylation was found in many loci of the immune system cells. But the most impressive results are coming from the Project Ice Storm cohort, which has been already mentioned as an example of prenatal maternal stress transmitted to offspring in Chapter 4.

Initially, it was noticed that children of mothers who were pregnant at the moment of the psychosocial stressful situation have demonstrated impairment of cognitive and language development in subsequent years and the level of impairment differed with respect to stress severity and trimester of pregnancy in which the disaster happened. The recent study concentrated on epigenetic markers widely distributed in the genome of immune cells of 13 years old adolescents, whose mothers have suffered the situation with the storm, but very importantly—taking into account such factor as a subjective rating of the stressful situation, that is, cognitive appraisal. It appeared that methylation of 2872 CpG pairs affiliated to 1564 genes differed significantly between adolescents from positive and negative maternal cognitive appraisal groups. Most of these genes belonged to biological pathways involved in the function of the immune system (Cao-Lei et al., 2015). These results suggest that pregnant women cognitive appraisal of a stressor, not only the stressor itself and even not only its' severity may have consequences for genome-wide methylation in the bodies and systems of their unborn children, which presumably persist for all their lives. So far as they have persisted for 13 years suggests that they will stay even longer, until time will come for their full maturation, the period of reproduction, working activity, and accumulation of usual stressors. Consequently, their impaired biological systems, metabolic or regulatory pathways, and immune system functions may result in different health effects and, quite probably, in their offspring too. It is quite possible that very soon, we will be able to see much more results of analogous studies.

This remarkable study helps to complete our model. It becomes clear that besides adverse event itself, subjective perception of it and the ability to cope (in full accordance with the definition of stress suggested by Lazarus) may have long-lasting transgenerational consequences with the participation of epigenetic mechanisms. In view of this, and taking into consideration all accumulated data, existing theories of suicide, epidemiological data on growth of suicides in adolescents, all theoretical and empirical data on growing psychosocial stress and allostatic load and considerations regarding evolution of behavior, an integrative model can be suggested that is based on understanding of the central role of epigenetic programming (Fig. 5.5). The model is conceptualized as "behavioral-psychological-existential" in accordance with main factors that are involved in multiple pathways leading to suicidal act. The model is conceived as holistic by nature, trying to avoid reductionism. According to the model, all interrelated events and interactions are developing under the general frame based on the most global processes associated with modernization—demographic changes, "compression of historical time," and loss of "reference points" and erosion of traditional values and "big meanings," as well as family crisis. It sets the frame for growing generalized stress and lowered possibility of social support in human population living in the modern society.

Next level of determinants is represented by existing macro- and mesofactors that determine health and disease, which are associated with prevailing socioeconomic system, according to the scheme presented in Fig. 5.1. Liberal economic model, which is prevailing in modern societies and makes them more or less safe from the point of view of development and sustainability, needs constant stimulation of consumption and, besides destroying ecological safety, impairs mental health of huge contingents by promoting in young people consumerism, hedonism, and unrealistic expectations. It is also associated with harsh inequalities, deprivation, and a variety of frustrations. These frustrations are usually associated with aggressive reactions, anxiety, and fears or pessimism about future. It is not consumption itself but comparisons with others actually leads to frustrations, hopelessness, and depressive symptoms, thus the main underlying factor is inequalities, which are growing worldwide and become much more obvious due to media activity. These factors determine the existing level of psychosocial stress, which is enhanced by work-related stress (lack of control and mismatch between efforts and rewards, time shifts, transport problem, etc.). This stress eventually turns into perceived subjective stress associated with anxiety, depressive thoughts, a negative vision of future, hopelessness, loss of meaning, and purpose in life (Fig. 5.5).

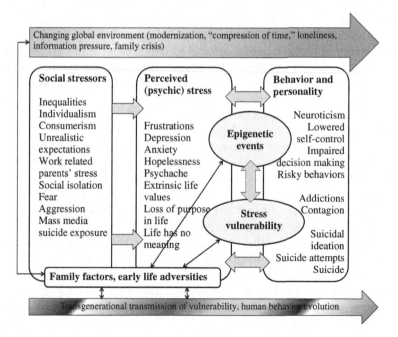

FIGURE 5.5 **Biobehavioral-psychological-existential model of suicide based on epigenetics.**

We consider that social stressors of mesolevel to a great extend play the initiating role in the whole scheme. Profound, but sometimes not easily acknowledged feelings and emotions that underlie perceived subjective stress are often hiding behind vague thoughts of being unsatisfied with life, frustrated, have not achieved desired goals, etc. This often happens to youngsters, who are entering "mature life" and feel a desire to capture a position in the social hierarchy but are often confronted by failures and frustrations due to unrealistic expectations and false understanding of life goals. Feeling of internal stress is equal to constant vigilance and being ready for action, so in a substantial part of young adults it is converted into a behavioral pattern of risky behaviors, early experimentation with psychoactive drugs and alcohol, promiscuity, and other signs of sensation seeking (allostatic behavior). It involves, besides internal physiological recursion, external behavioral recursion, and leads to finding a social niche (reference group, specific environment, etc.) that supports such behaviors (Cole, 2013). Such situation is rather typical for adolescents who may be eager to prove themselves, to be successful at least in something, including dubious achievements, or to find understanding of their problems and recognition in a group. If such group is a social media suicide enthusiasts' network or a group of adepts of a destructive cult, the risk grows immensely. Entering such group may be the example of rGE, that is, active covariation between genes and environment.

Modern approaches to understanding the role of stress in an individual life are based on integrative, holistic thinking. Epigenetics have brought much better understanding how social factors actually influence behaviors and mental health, vulnerability, or resilience. Stress in parents generates a response of the parental organism, which determines vulnerability to the new stress in the offspring by epigenetic programming of behaviors and traits. If stress in a couple of decades is enhanced, it will lead to the differential reactivity of those, whose personality traits and behavioral responses are matching (or not matching) new social and other environmental conditions. It may lead both to adaptive or maladaptive responses, depending on many factors, genetic and epigenetic by nature (Del Giudice, 2012). We suggest that it will lead, given that most of the traits and behaviors are normally distributed in the population, to higher representation of extreme qualities that may be found on both sides of the curve. There is no doubt that some individuals will benefit from stressful environments, with information overload, inequalities and dynamic changes in all spheres of life. On the other hand, the portion of those who will develop mental health problems and behavioral abnormalities will also grow. Such thinking is in concordance with existing models that put forward epigenetic programming and environmental match/mismatch, and stress accumulation model, which is very applicable for the period of maturation of young adults. As Nederhof and Schmidt (2012) have proposed, mismatch hypothesis may be applicable to those individuals, who experienced strong programming in early childhood, while cumulative hypothesis explains behaviors of those who avoided harsh programming. This remark, to our opinion, reconciles "structural" model of Turecki et al. (2012) and our model. Another interesting confirmation comes from field studies by Belsky and Beaver (2011) who have pointed to a possible role of the cumulative effect of the so-called "plasticity alleles" in responsiveness of children to parenting influences, both negative and positive. Among plasticity alleles, there are markers of dopaminergic and serotoninergic systems known as predictors of some personality traits, conduct disorders, and depression (Belsky & Beaver, 2011; Beaver & Belsky, 2012). The authors conceptualize their hypothesis as differential susceptibility, which due transgenerational transmission of parenting style can explain progressions both "to the worth and to the better." But it should be understood not as an isolated mechanism, all interactions happen in a specific social situation and effects of the environment may have much bigger influence than genetic predispositions. It may be very relevant for suicide growth.

Suicide in this context is one of the consequences of growing pressure of stress and lowered social support, bad parenting, and it's transgenerational transmission, interaction with the environment being one of the most potent factors in this complex scheme. As Curtis, Curtis, and Fleet (2013) have noticed, each generation of young people, who are going to be parents, enter adult life with high expectations and each age cohort meets their own deprivations due to environmental socioeconomic conditions. In changing generations, it may lead to an evolution of personality traits, for instance, neuroticism, less self-control, risky and unhealthy behaviors, suicidal thoughts, attempts, and completed suicide. We have been discussing the growth of neuroticism scores in young people cohorts for the last 50–60 years (Chapter 1). Neuroticism is the factor of psychopathologies and is enhanced in suicide attempters; moreover, it has its own dynamics during maturation. In particular, it transition from adolescence to young adulthood individuals with a higher absolute level of neuroticism were at 14-fold increased the risk for depression and 7-fold risk for anxiety disorders at the age of 25 (Aldinger et al., 2014). Thus, a period of life when first frustrations will happen (which to a certain extent depends on personality, genetics, and epigenetics, associated with stress-vulnerability) may also have serious consequences. It is also very important if reproduction will coincide with such vulnerable period or not. Risk factors for suicide in a cohort study, except low birth weight and high order of birth (i.e., being the last child in a big family), were also too young and not well-educated mothers, early motherhood being one of the strongest predictors (Mittendorfer-Rutz, Rasmussen, & Wasserman, 2004). It fits fairly well all models, including those that focus on early life stress and bad parenting, and those that focus on stress-vulnerability and stress accumulation in the period of transition from adolescence to adulthood.

Changing representation of personality traits and behaviors add to existing stress especially in cases when supportive resources are scarce or not available for the vulnerable individual. Epigenetic mechanisms in these interactions serve the role of interface between social environment, internal subjective stressful feelings, emotions and experiences, and constantly changing genome. Thus, as it was conceptualized by Meloni (2014), social brain meets reactive genome. Epigenome in this sense appears in the role of the mechanism of embedding of social experiences and their further transmission in generations, genes being not the "leaders," but the "followers" of the evolution of human behavior.

Such schematization includes, or at least tries to include all types of interactions and possible reasons for suicide, especially with regards of heterogeneity of suicides. Early life adversities and their epigenetic embedding, leading to higher stress-vulnerability due to structural changes emerging in development, may play a decisive role in some smaller part of the suicidal youngsters, having such features as depression, anxiety, behavioral abnormalities, and cognitive impairment, that is, poor problem solving and cognitive rigidity. Later in life emerging sensitivity and consequent frustrations, together with the loss of purpose of life, idleness, virtualization of social contacts, narcissism, low self-perception, and negativism, dependence on the opinion of a group, identification with a suicide of the significant other may lead to suicidal act in another part of young people. This part is obviously bigger and presumably does not belong to the vulnerable subpopulation with low socioeconomic status, multiple environmental risks, and other peculiarities which make it easily detectable and can be targeted by preventive strategies. It makes the task of prevention more complicated.

So, it becomes clear that experience of earlier social events, ability to cope with the threats and perceived subjective feeling of stress are incorporated into biological systems in the form of epigenetic marks. An infinite number of double-way causal relationships and loops between the society, the individual, personal experiences, memory, and biological systems create a complex picture of interactions. The leading role belongs to the increasing pressure of the environment in the form of psychosocial stress. The environment appears to be a more potent factor that it was perceived before. This pressure results in the evolution of personality traits and behaviors, which are risk factors for psychopathologies and suicide. This may lead to mental disorders, such as depression, anxiety, and addictions, which are also known as high-risk factors for suicide. Such conceptualization is supported by opinions of many authors, who consider that environmental influences are the primary causal factor that shapes transcriptional response of neural and endocrine systems thus programming development of the disorder (Szyf, 2011; McEwen, 2012; Lester et al., 2012).

Presented model suggests several cooperated and interrelated pathways and mechanisms. Each pathway and mechanism may add a portion to generally measured outcome—suicides in a younger generation. Suicide is a complex and multifaceted and heterogeneous phenomenon, which has many reasons, mechanisms, and risk factors. Analysis of existing sources gives a clear impression that epigenetics can be viewed as a central mechanism that unites and integrates many risk factors of suicide and which has a potential to explain the growth of suicides in the young people worldwide.

References

Abdoplmaleky, H. M., Yagubi, S., Papageorgis, P., Lambert, A., Ozturk, S., Sivaraman, V., & Thiagalingam, S. (2011). Epigenetic dysregulation of 5-HTR2A in the brain of patients with schizophrenia and bipolar disorder. *Schizophrenia Research, 129*, 183–190.

Agerbo, E., Nordentoft, M., & Mortensen, P. B. (2002). Familial, psychiatric, and socioeconomic risk factors for suicide in young people: nested case-control study. *British Medical Journal, 325*, 74.

Aldinger, M., Stopsack, M., Ulrich, I., Appel, K., Reinelt, E., Wolff, S., Grabe, H. J., Lang, S., & Barnow, S. (2014). Neuroticism developmental courses—implications for depression, anxiety and everyday emotional experience; a prospective study from adolescence to young adulthood. *BMC Psychiatry, 14*, 210.

Apter, A., & Gvion, Y. (2016). Adolescent suicide and attempted suicide. In D. Wasserman (Ed.), *Suicide. An unnecessary death* (2nd ed., pp. 197–213). New York: Oxford University Press.

Arango, V., Huang, Y., Underwood, M. D., & Mann, J. J. (2002). Genetics of the serotoninergic system in suicidal behavior. *Journal of Psychiatric Research, 37*, 375–386.

Asberg, M. (1997). Neurotransmitters and suicidal behavior. The evidence form cerebrospinal fluid studies. *Annals of the New York Academy of Science, 836*, 158–181.

Baldessarini, R. J., & Hennen, J. (2004). Genetics of suicide: an overview. *Harvard Review of Psychiatry, 12*, 1–13.

Bale, T. (2014). Lifetime stress experiences: transgenerational epigenetics and germ cells programming. *Dialogues in Clinical Neuroscience, 16*, 297–305.

Bani-Fatemi, A., Howe, A. S., & De Luca, V. (2015). Epigenetic studies of suicidal behavior. *Neurocase, 21*, 134–143.

Batchelor, I., & Napier, M. (1953). Attempted suicide in the old age. *British Medical Journal, 2*, 1186–1190.

Baumeister, R. F. (1990). Suicide as escape from self. *Psychological Review, 97*, 90–113.

Beaver, K. M., & Belsky, J. (2012). Gene-environment interaction and the intergenerational transmission of parenting: testing the differential-susceptibility hypothesis. *Psychiatric Quarterly, 83*, 29–40.

Beck, A. T., Brown, G., Berchick, R., Stewart, B. L., & Steer, R. A. (1990). Relationship between hopelessness and ultimate suicide: a replication with psychiatric outpatients. *American Journal of Psychiatry, 147*, 190–195.

Belsky, J., & Beaver, K. M. (2011). Cumulative-genetic plasticity, parenting and adolescent self-regulation. *Journal of Child Psychology and Psychiatry, 52*, 619–626.

Binder, E. B., Bradley, R. G., Liu, W., Epstein, M. P., Deveau, T. C., Mercer, K. B., Tang, Y., Gillespie, C. F., Heim, C. M., Nemeroff, C. B., Schwartz, A. C., Cubells, J. F., & Ressler, K. J. (2008). Association of FKBP5 polymorphisms and childhood abuse with risk of post-traumatic stress disorder symptoms in adults. *Journal of American Medical Association, 299*, 1291–1305.

Bodner, E., Ben-Artzi, E., & Kaplan, Z. (2006). Soldiers who kill themselves: the contribution of dispositional and situational factors. *Archives of Suicide Research, 10*, 29–43.

Bondy, B., Buettner, A., & Zill, P. (2006). Genetics of suicide. *Molecular Psychiatry, 11*, 336–351.

Brent, D. A., & Mann, J. J. (2005). Family genetic studies, suicide, and suicidal behavior. *American Journal of Medical Genetics, 133C*, 13–24.

Brent, D. A., & Melhem, N. (2008). Familial transmission of suicidal behavior. *Psychiatric Clinics of North America, 31*, 157–177.

Brent, D. A., Bridge, J., Johnson, B. A., & Connolly, J. (1996). Suicidal behavior runs in families. A controlled study of adolescent suicide victim. *Archives of General Psychiatry, 53*, 1145–1152.

Brezo, J., Bureau, A., Mérette, C., Jomphe, V., Barker, E. D., Vitaro, F., Hébert, M., Carbonneau, R., Tremblay, R. E., & Turecki, G. (2010). Differences and similarities in the serotonergic diathesis for suicide attempts and mood disorders: a 22-year longitudinal gene-environment study. *Molecular Psychiatry, 15*, 831–843.

Cacioppo, J. T., Fowler, J. H., & Christakis, N. A. (2009). Alone in the crowd: the structure and spread of loneliness in a large social network. *Journal of Personal and Social Psychology, 97*, 977–991.

Cao-Lei, L., Elgbeili, G., Massart, R., Laplante, D. P., Szyf, M., & Kimg, S. (2015). Pregnant women cognitive appraisal of a natural disaster affects DNA methylation in their children 13 years later: project Ice Storm. *Translational Psychiatry, 5*, e515.

Caspi, A., Sugden, K., Moffitt, T. E., Taylor, A., Craig, I. W., Harrington, H. -L., McClay, J., Mill, J., Martin, J., Braithwaite, A., & Poulton, R. (2003). Influence of life stress on depression: moderation by a polymorphism in the 5-HTT gene. *Science, 301*, 386–389.

Cleveland, H. H., Wiebe, R. P., & Rowe, D. C. (2005). Sources of exposure to smoking and drinking friends among adolescents: a behavioral-genetic evaluation. *The Journal of Genetic Psychology, 166*, 153–169.

Cole, S. W. (2013). Social regulation of human gene expression: mechanisms and implications for public health. *American Journal of Public Health, 103*(Suppl. 1), S84–S92.

Corral-Frías, N. S., Nikolova, Y. S., Michalski, L. J., Baranger, D. A., Hariri, A. R., & Bogdan, R. (2015). Stress-related anhedonia is associated with ventral striatum reactivity to reward and transdiagnostic psychiatric symptomatology. *Psychological Medicine, 45*, 2605–2617.

Crews, D. (2010). Epigenetics, brain, behavior, and the environment. *Hormones, 9*, 41–50.

Curley, J. P., & Champaigne, F. A. (2016). Influence of maternal care on the developing brain: mechanisms, temporal dynamics and sensitive periods. *Frontiers in Neuroendocrinology, 40*, 52–66.

Curtis, B., Curtis, C., & Fleet, R. (2013). Socio-economic factors and suicide: the importance of inequality. *New Zealand Sociology, 28*, 77–92.

Del Giudice, M. (2012). Early stress and human behavioral development: emerging evolutionary perspectives. *Journal of Developmental Origin of Health and Disease, 5*, 270–280.

DeLuca, V., Viggiano, E., Dhoot, R., Kennedy, J. L., & Wong, A. H. (2009). Methylation and QTDT analysis of the 5-HT2A receptor 102C allele: analysis of suicidality in major psychosis. *Journal of Psychiatric Research, 43*, 532–537.

Durkheim, E. (1897). *Le suicide: etude de sociologie.* Paris: Felix Alcan.

Egeland, J. A., & Sussex, J. N. (1985). Suicide and family loading for affective disorders. *JAMA: the Journal of the American Medical Association, 254*, 915–918.

El Ansari, W., Oskrochi, R., & Haghoo, G. (2014). Are students' symptoms and health complaints associated with percieved stress at university? Perspectives from the United Kingdom and Egypt. *International Journal of Environmental Research and Public Health, 11*, 9981–10002.

El-Sayed, A. M., Haloossim, M. R., Galea, S., & Koenen, K. (2012). Epigenetic modifications associated with suicide and common mood and anxiety disorders: a review of the literature. *Biology of Mood and Anxiety Disorders, 2*, 10.

Engel, G. L. (1997). The need for a new medical model: a challenge for biomedicine. *Science, 196*, 129–136.

Ernst, C., Deleva, V., Deng, X., Sequeira, A., Pomarenski, A., Klempan, T., Ernst, N., Quirion, R., Gratton, A., Szyf, M., & Turecki, G. (2009). Alternative splicing, methylation state, and expression profile of tropomyosin-related kinase B in the frontal cortex of suicide completers. *Archives of General Psychiatry, 66*, 22–32.

Farber, M. L. (1968). *Theory of suicide.* New York: Funk & Wagnalls.

Farberow, N. L. (Ed.). (1980). *The many faces of suicide.* New York: McGraw-Hill Book Company.

Fergusson, D. M., Woodward, L. J., & Horwood, L. J. (2000). Risk factors and life processes associated with the onset of the suicidal behavior during adolescence and early adulthood. *Psychological Medicine, 30*, 23–39.

Finkelhor, D., Shattuck, A., Turner, H. A., & Hamby, S. L. (2014). Trends in children's exposure to violence, 2003 to 2011. *JAMA Pediatrics, 168*, 540–546.

Flint, J., & Munafo, M. R. (2007). The endophenotype concept in psychiatric genetics. *Psychological Medicine, 37*, 163–180.

Garfinkel, B., Froese, A., & Hood, J. (1982). Suicide attempts in children and adolescents. *American Journal of Psychiatry, 139*, 1257–1261.

Garner, A. S. (2013). Home visiting and the biology of toxic stress: opportunities to address early childhood adversity. *Pediatrics, 132*(Suppl. 2), S65–S73.

Gilger, J. W. (2000). Contribution and promise of human behavioral genetics. *Human Biolology, 72*, 229–255.

Glick, A. R. (2015). The role of serotonin in impulsive aggression, suicide, and homicide in adolescents and adults: a literature review. *International Journal of Adolescents Mental Health, 27*, 143–150.

Goldney, R. (2000). Ethology and suicidal behavior. In K. Hawton, & K. van Heeringen (Eds.), *The international handbook of suicide and attempted suicide (94–106).* Chichester: John Wiley & Sons.

Golubovskii, M. D. (2000). *Century of genetics: Evolution of ideas and concepts.* St. Petersburg: Borei Art.

Gross, J. A., Fiori, L. M., Labonté, B., Lopez, J. P., & Turecki, G. (2013). Effects of promoter methylation on increased expression of polyamine biosynthetic genes in suicide. *Journal of Psychiatric Research, 47*, 513–519.

Haghighi, F., Xin, Y., Chanrion, B., O'Donnel, A. H., Ge, Y., Dwork, A. J., Arabgo, V., & Mann, J. J. (2014). Increased DNA methylation in the suicide brain. *Dialogues in Clinical Neuroscience, 16*, 430–438.

Hochberg, Z., Feil, R., Constancia, M., Fraga, M., Junien, C., Carel, J. -C., Boileau, P., Le Bouc, Y., Deal, C. L., Lillycrop, K., Scharfmann, R., Sheppard, A., Skinner, M., Szyf, M., Waterland, R. A., Waxmann, D. J., Whitelaw, E., Ong, K., & Albertson-Wikland, K. (2011). Child health, developmental plasticity, and epigenetic programming. *Endocrine Reviews, 32*, 159–224.

Joiner, T. E., Brown, J. S., & Wingate, L. R. (2005). The psychology and neurobiology of suicidal behavior. *Annual Review of Psychology, 56*, 287–314.

Kaminsky, Z., Wilcox, H. C., Eaton, W. W., Van Eck, K., Kilaru, V., Jovanovic, T., Klengel, T., Bradley, B., Binder, E. B., Ressler, K. J., & Smith, A. K. (2015). Epigenetic and genetic variations at SKA2 predict suicidal behavior and post-traumatic stress disorder. *Translational Psychiatry, 5*, e627.

Kang, H. -J., Kim, J. -M., Lee, J. -Y., Kim, S. -Y., Bae, K. -Y., Kim, S. -W., Shin, I. -S., Kim, H. -R., Shin, M. -G., & Yoon, J. -S. (2013). BDNF promoter methylation and suicidal behavior in depressive patients. *Journal of Affective Disorders, 151*, 679–685.

Keller, S., Sarchiapone, M., Zarrilli, F., Videtic, A., Ferraro, A., Carli, V., Sacchetti, S., Lembo, F., Angiolillo, A., Jovanovic, L., Pisanti, F., Tomaiuolo, R., Monticelli, A., Balazic, J., Roy, A., Marusic, A., Cocozza, S., Fusco, A., Bruni, C. B., Castaldo, G., & Chariotti, L. (2012). Increased BDNF promoter methylation in the Wernicke area of suicide subjects. *Archives of General Psychiatry, 69*, 62–70.

Kendler, K. S. (2010). Genetic and environmental pathways to suicidal behavior: reflections of a genetic epidemiologist. *European Psychiatry, 25*, 300–303.

Kendler, K. S., Neale, M., Kessler, R., Heath, A., & Eaves, L. (1993). A twin study of recent life events and difficulties. *Archives of General Psychiatry, 50*, 789–796.

Keverne, E. B., & Curley, J. P. (2008). Epigenetics, brain evolution and behavior. *Frontiers of Neuroendocrinology, 29*, 398–412.

Kiekens, G., Bruffaerts, R., Nock, M. K., Van de Ven, M., Witteman, C., Mortier, P., Demyttenaere, K., & Claes, L. (2015). Non-suicidal: self-injury among Dutch and Belgian adolescents: personality, stress and coping. *European Psychiatry, 30*, 743–749.

Kokonei, G., Jozan, A., Morgan, A., Szemenei, E., Urban, R., Reinhardt, M., & Demetrovics, Z. (2015). Perseverative thoughts and subjective health complaints in adolescence: mediating effects of perceived stress and negative effect. *Psychology & Health, 30*, 969–986.

Kubota, T., Miyake, K., Hariya, N., & Mochizuki, K. (2015). Understanding the epigenetics of neurodevelopmental disorders and DOHaD. *Journal of Developmental Origins of Health and Disease, 6*, 96–104.

Labonte, B., & Turecki, G. (2010). The epigenetics of suicide: explaining the biological effect of early life environmental adversity. *Archives of Suicide Research, 14*, 291–310.

Lahey, B., Moffitt, T. E., & Caspi, A. (Eds.). (2003). *Causes of conduct disorder and juvenile delinquency.* New York: Guilford Press.

Lam, L. L., Emberly, E., Fraser, H. B., Neumann, S. M., Chen, E., Miller, G. E., & Kobor, M. S. (2012). Factors underlying variable DNA methylation in a human community cohort. *Proceeding of the National Academy of Sciences of USA, 109*(Suppl. 2), 17253–17260.

Le-Nicilesku, A. B., Levey, D., Le-Nikulesku, H., Ayalew, M., Palmer, L., Gavrin, L. M., Jain, N., Winiger, E., Bhosrekar, S., Shankar, G., Radel, M., Bellanger, E., Duckworth, H., Olesek, K., Vergo, J., Schweitzer, R., Yard, M., Ballew, A., Shekhar, A., Sandusky, G. E., Schork, N. J., Kurian, S. M., Salomon, D. R., & Niculescu, A. B., 3rd (2013). Discovery and validation of blood biomarkers for suicidality. *Molecular Psychiatry, 164*, 118–122.

Lester, B. M., Marsit, C. J., Conradt, E., Bromer, C., & Padbury, J. F. (2012). Behavioral epigenetics and the developmental origin of child mental health. *Journal of Developmental Origins of health and disease, 3*, 395–408.

Lin, P. Y., & Tsai, G. (2004). Association between serotonin transporter gene polymorphism and suicide: results of a meta-analysis. *Biological Psychiatry, 55*, 1023–1030.

Lockwood, L. E., Su, S., & Youssef, N. A. (2015). The role of epigenetics in depression and suicide: A platform for gene-environment interactions. *Psychiatry Research, 228*, 235–242.

Lorenzo-Blanco, E. I., & Unger, J. B. (2015). Ethnic discrimination, acculturative stress, and family conflict as predictors of depressive symptoms and cigarette smoking among Latina/o youth: the mediating role of perceived stress. *Journal of Youth and Adolescence, 44*, 1984–1997.

Luijcks, R., Vossen, C. J., Hermens, H. J., van Os, J., & Lousberg, R. (2015). The influence of percieved stress on cortical reactivity: a proof-of-principle study. *PLOS One, 10*(6), e0129220.

Mann, J. J. (1998). The neurobiology of suicide. *Nature Medicine, 4*, 25–30.

Mann, J. J., Malone, K. M., Nielsen, D., Goldman, D., Erdos, J., & Gelernter, J. (1997). Possible association of a polymorphism of a tryptophan hydroxylase gene with suicidal behavior in depressed patients. *American Journal of Psychiatry, 154*, 1452–1453.

Mann, J. J., Huang, Y., Underwood, M. D., Kassir, S. A., Oppenheim, S., Kelly, T. M., Dwork, A. J., & Arango, V. (2000). A serotonin transporter gene promoter polymorphism (5-HTTLPR) and prefrontal cortical binding in major depression and suicide. *Archives of General Psychiatry, 57*, 729–738.

Mansell, T., Novakovic, B., Meyer, B., Rzehak, P., Vuillermin, P., Ponsonby, A. -L., Collier, F., Burgner, D., Saffery, R., & Ryan, J. & BIS investigator team. (2016). The effects of maternal anxiety during pregnancy on *IGF2/H19* methylation in cord blood. *Translational Psychiatry, 6*, e765.

Manteaux, P., Jacques, D., & Zdanovich, N. (2015). Hormonal and developmental influences on adolescent suicide: a systematic review. *Psychiatria Danubina*, 27(Suppl. 1), 300–304.

McEwen, B. (2012). Brain on stress: How the social environment gets under the skin. *Proceedings of National Academy of Sciences*, 109(Suppl. 2), 17180–17185.

McEwen, B., Gray, J. D., & Nasca, C. (2015). Recognizing resilience: learning from the effects of stress on the brain. *Neurobiology of Stress*, 1, 1–11.

McGowan, P. O., Sasaki, A., Huang, T., Unterberger, A., Suderman, M., Ernst, C., Meaney, M. J., Turecki, G., & Szyf, M. (2008). Promoter-wide hypermethylation of the ribosomal RNA gene promoter in the suicide brain. *PLOS One*, 3, e2085.

McGowan, P. O., Sasaki, A., D'Alessio, A. C., Dymov, S., Labonté, B., Szyf, M., Turecki, G., & Meaney, M. J. (2009). Epigenetic regulation of the glucocorticoid receptor in human brain associates with childhood abuse. *Nature Neuroscience*, 12, 342–348.

McKeown, R. E., Cuffe, S. P., & Schulz, R. M. (2006). US suicide rates by age group, 1970–2002: an examination of recent trends. *American Journal of Public Health*, 96(10), 1744–1751.

Meloni, M. (2014). The social brain meets the reactive genome: neuroscience, epigenetics and the new social biology. *Frontiers in Human Neuroscience*, 8, 309.

Mittendorfer-Rutz, E., Rasmussen, F., & Wasserman, D. (2004). Restricted fetal growth and adverse maternal psychosocialand socioeconomic conditions as risk factors for suicidal-behaviour of offspring: a cohort study. *Lancet*, 364, 1135–1140.

Moffitt, T. E. (2005). Genetic and environmental influences in antisocial behaviors: evidence from behavioral-genetic research. *Advanced Genetics*, 55, 41–104.

Morin-Major, J. K., Marin, M. F., Durand, N., Wan, N., Juster, R. P., & Lupien, S. J. (2015). Facebook behaviors associated with diurnal cortisol in adolecents: is befriending stressful? *Psychoneuroendocrinology*, 63, 238–246.

Murdock, K. K., Gorman, S., & Robbins, M. (2015). Co-rumination via cellphone moderates the association of perceived interpersonal stress and psychoscial well-being in emerging adults. *Journal of Adolescence*, 38, 27–37.

Murphy, G., & Wetzel, R. D. (1982). Family history of suicidal behavior among suicide attempters. *The Journal of Nervous and Mental Disease*, 180, 86–90.

Murphy, G. E., Wetzell, R. D., Swallow, C. S., & McClure, J. N. (1969). Who calls the suicide prevention center: a study of 55 persons calling on their own behalf. *American Journal of Psychiatry*, 126, 313–324.

Nagy, C., & Turecki, G. (2012). Sensitive periods in epigenetics: bringing us closer to complex behavioral phenotypes. *Epigenomics*, 4, 445–457.

Nagy, C., Suderman, M., Yang, J., Szyf, M., Machawar, N., Ernst, C., & Turecki, G. (2015). Astrocytic abnormalities and global DNA methylation patterns in depression and suicide. *Molecular Psychiatry*, 20, 320–328.

Nazarov, V. I. (2007). *Evolutsiya ne po Darwinu. Smena evolutsionnoy modeli. (Evolution not according to Darwin. Change of the evolutionary model)* (2nd ed.). Moscow: LKI.

Nederhof, E., & Schmidt, M. V. (2012). Mismatch or cumulative stress: toward an integrated hypothesis of programming effects. *Physiology & Behavior*, 106, 691–700.

Nemoda, Z., Massart, R., Suderman, M., Hallett, M., Li, T., Coote, M., Cody, N., Sun, Z. S., Soares, C. N., Turecki, G., Steiner, M., & Szyf, M. (2015). Maternal depression is associated with DNA methylation changes in cord blood T lymphocytes and adult hippocampi. *Translational Psychiatry*, 5, e545.

Nieto, S. J., Patriquin, M. A., Nielsen, D. A., & Kosten, T. A. (2016). Don't worry; be aware of the epigenetics of anxiety. *Pharmacolology Biochemistry and Behavior*, 146-147, 60–72.

O'Connor, R. C., & Nock, M. K. (2014). The psychology of suicidal behavior. *The Lancet*, 1, 73–85.

Orbach, I. (1994). Dissociation, physical pain, and suicide: a hypothesis. *Suicide and Life-Threatening Behavior*, 24(1), 68–79.

Pantic, I. (2014). Online social networking and mental health. *Cyberpsychology, Behavior and Social Networking*, 17, 652–657.

Park, S. (2014). Association of physical activity, with sleep satisfaction, perceived stress, and problematic Internet use in Korean adolescents. *BMC Public Health, 14,* 1143.

Perroud, N., Courtet, P., Vincze, I., Jaussent, I., Jollant, F., Bellivier, F., Leboyer, M., Baud, P., Buresi, C., & Malafosse, A. (2008). Interaction between BDNF Val66Met and childhood trauma on adult's violent suicide attempt. *Genes, Brain and Behavior, 7,* 314–322.

Pfeffer, C. R. (2000). Suicidal behavior in children: an emphasis on developmental influences. In K. Hawton, & K. van Heeringen (Eds.), *The international handbook of suicide and attempted suicide* (pp. 237–248). Chichester: John Wiley & Sons.

Pilgrim, D. (2002). The biopsychosocial model in Anglo-American psychiatry: past, present and future. *Journal of Mental Health, 11,* 585–594.

Poulter, M. D., Du, L., Weaver, I. C. G., Palkovits, M., Faludi, G., Merali, Z., Szyf, M., & Anisman, H. (2008). GABAA receptor promoter hypermethylation in suicide brain: implication for the involvement of epigenetic processes. *Biological Psychiatry, 64,* 645–652.

Qin, P., Agerbo, E., & Mortensen, P. B. (2002). Suicide risk in relation to family history of completed suicide and psychiatric disorders: a nested case-control study based on longitudinal registers. *Lancet, 360,* 1126–1130.

Qiu, A., Anh, T. T., Li, Y., Chen, H., Rifkin-Graboi, A., Broekman, B. F. P., Kwek, K., Saw, S. -M., Chong, Y. -S., Gluckman, P. D., Fortier, M. V., & Meaney, M. J. (2015). Prenatal maternal depression alters amygdala functional connectivity in 6-month-old infants. *Translational Psychiatry, 5,* e508.

Rakhimkulova, A.V. & Rozanov V.A. (2015). Perceived stress, anxiety, depression and risky behavior in adolescents. In *22nd Multidisciplinary ISBS International Neuroscience and Biological Psychiatry "Stress and Behavior"* p. 31–32. S.Petersburg.

Rangaraju, S., Levery, D. F., Nho, K., Jain, N., Andrews, K. D., Le-Niculescu, H., Salomon, D. R., Saykin, A. J., Petrascheck, M., & Niculescu, A. B. (2016). Mood, stress and longevity: convergence on ANK3. *Molecular Psychiatry, 21,* 1037–1049.

Reiss, D., Hetherington, E. M., Plomin, R., Howe, G. W., Simmens, S. J., Henderson, S. H., O'Connor, T. J., Bussell, D. A., Anderson, E. R., & Law, T. (1995). Genetic questions for environmental studies. Differential parenting and psychopathology in adolescences. *Archives of General Psychiatry, 52,* 925–936.

Robins, E., Schmidt, E. H., & O'Neal, P. (1957). Some interrelation of social factors and clinical diagnosis in attempted suicide: a study of 109 patients. *American Journal of Psychiatry, 114,* 221–231.

Romeo, R. D. (2010). Adolescence: a central event in shaping stress reactivity. *Developmental Psychobiology, 52,* 244–253.

Rowe, D. C. (2003). Assessing genotype-environment interactions and correlations in the postgenomic era. In R. Plomin, J. DeFries, I. W. Craig, & P. McGuffin (Eds.), *Behavioral genetics in the post-genomic era* (pp. 71–86). Washington, DC: American Psychological Association.

Roy, A. (1983). A family history of suicide. *Archives of General Psychiatry, 40,* 971–974.

Roy, A. (1993). Genetic and biologic risk factors for suicide in depressive disorders. *Psychiatric Quarterly, 64,* 345–358.

Roy, A., Hu, Z. X., Janal, M. N., & Goldman, D. (2007). Interaction between childhood trauma and serotonin transporter gene variation in suicide. *Neuropsychopharmacology, 32,* 2046–2052.

Roy, A., Gorodetsky, E., Yuan, G., Goldman, D., & Enoch, M. -A. (2010). Interaction of FKBP5, a stress-related gene, with childhood trauma increases the risk for attempting suicide. *Neuropsychopharmacology, 35,* 1674–1683.

Roy, A., Segal, N., Centerwall, D., & Robinette, C. D. (1991). Suicide in twins. *Archives of Genetic Psychiatry, 48,* 29–32.

Roy, A., Segal, N. L., & Sarchiapore, M. (1995). Attempted suicide among living co-twins of twin suicide victims. *American Journal of Psychiatry, 152,* 1075–1076.

Roy, A., Rylander, G., & Sarchiapone, M. (1997). Genetic studies of suicidal behavior. *The Psychiatric Clinics of North America, 20,* 595–611.

Rozanov, V. A. (2010). O mechanismas suitsidal'nogo povedeniya i vosmozhnostiah ego prediktsii na rannyh etapah razvitiya (Mechanisms of suicidal behavior and perspectives of its' prediction on the early stages of development). *Ukrainian Medical Journal, 75,* 92–97.

Rozanov, V. A. (2013). Geny i suitsidal'noye povedeniye (Genes and suicidal behavior). *Suitsidologia (Suicidology), 4,* 3–14.

Rozanov, V. A. (2014). Suitsidy sredy podrostkov—chto proishodit i v chem prichina? (Suicides in children and adolescents—what is happening and what may be the reason?). *Suitsidologia (Suicidology), 5,* 16–31.

Rozanov, V. A. (2015). Stress-indutsirovannie epigeneticheskie fenomeny—esche odin veroyatniy biologichesly factor suitsida (Stress-induced epigenetic phenomena—one more biological mechanism of suicide). *Sutsidologia (Suicidology), 6,* 3–19.

Rozanov, V. A., & Carli, V. (2012). Suicide among War Veterans. *International Journal of Environmental Research and Public Health, 9,* 2504–2519.

Rozanov, V. A., & Mid'ko, A. A. (2006). Systemic lipid metabolism and suicidal behavior (Systemniy lipidniy obmen I suitsidal'noe povedenie). *Neyronauki (Neurociences), 4,* 3–13.

Rozanov, V. A., Mokhovikov, A. N., & Wasserman, D. (1999). Neurobiological basis of suicidality. *Ukrainian Medical Journal, 6,* 5–12.

Rozanov, V. A., Yemyasheva, Zh. V., & Biron, B. V. (2011). Vliyanie travmy detskogo vozrasta na nakoplenie stressovych sobytiy i formirovanie suitsidal'nych tendentsiy v techenie zhizny (Influence of childhood traumatic experience on general life stress accumulation and suicidal tendencies through the life-span). *Ukrainian Medical Journal, 6,* 94–98.

Rudd, M. D. (2000). The suicidal mode: a cognitive-behavioral model of suicidality. *Suicide and Life-Threatening Behavior, 30,* 18–33.

Rudd, M. D. (2006). Fluid vulnerability theory: a cognitive approach to understanding the process of acute and chronic suicide risk. In P. T. Ellis (Ed.), *Cognition and suicide: Theory, research and therapy* (pp. 355–368). American Psychological Association: Washington, DC.

Rudd, M. D., Trotter, D. R. M., & Williams, B. (2009). Psychological theories of suicide. In D. Wasserman, & C. Wasserman (Eds.), *Oxford textbook on suicidology and suicide prevention* (pp. 159–164). NY: Oxford University Press.

Rujesku, D., Thalmeier, A., Moller, H. -J., Bronisch, T., & Giegling, I. (2007). Molecular genetics findings in suicidal behavior: what is beyond the serotoninegric system? *Archives of Suicide Research, 11,* 17–40.

Sarchiapone, M., & Iosue, M. (2015). Genetics of suicidal behavior. In U. Kumar (Ed.), *Suicidal behavior. Underlying dynamics* (pp. 24–38). New York, London: Routledge.

Schneider, E., El Hajj, N., Müller, F., Navarro, B., & Haaf, T. (2015). Epigenetic dysregulation in the prefrontal cortex of suicide completers. *Cytogenetic and Genome Research, 146,* 19–27.

Schneidman, E. S. (2001). *Comprehending suicide: Landmarks in 20th-century suicidology.* Washington, DC: American Psychological Association.

Schützenberger, A. A. (1998). *The ancestor syndrome.* London & New York: Routledge.

Segal, N. L., & Roy, A. (1995). Suicide attempts in twins whose co-twins deaths were not suicides. *Personality and Individual Differences, 19,* 937–940.

Selby, E. A., Anestis, M. D., Bender, T. W., Ribeiro, J. D., Nock, M. K., Rudd, M. D., Bryan, C. J., Lim, I. C., Baker, M. T., Gutierrez, P. M., & Joiner, T. E., Jr. (2010). Overcoming the fear of lethal injury: evaluating suicidal behavior in the military through the lens of the interpersonal–psychological theory of suicide. *Clinical Psychology Reviews, 30,* 298–307.

Shaw, D. M., Camps, F. E., & Eccleston, E. G. (1967). 5-Hydroxytryptamine in the hindbrain of depressive suicides. *British Journal of Psychiatry, 113,* 1407–1411.

Short, A. K., Fennell, K. A., Perreau, V. M., Fox, A., O'Bryan, M. K., Kim, J. H., Bredy, T. W., Pang, T. Y., & Hannan, A. J. (2016). Elevated paternal glucocorticoid exposure alters the small noncoding RNA profile in sperm and modifies anxiety and depressive phenotypes in the offspring. *Translational Psychiatry, 6,* e837.

Shulsinger, F., Kety, S., Rosenthal, D., & Wender, P. (1979). A family story of suicide. In M. Schou, & E. Stromgen (Eds.), *Origins, prevention and treatment of affective disorders* (pp. 277–278). NY: Academic Press.

Statham, D. J., Heath, A. C., Madden, P. A. F., Bucholz, K. K., Beirut, L., Dinwiddie, S. H., Slutske, W. S., Dunne, M. P., & Martin, N. G. (1998). Suicidal behavior: an epidemiological and genetic study. *Psychological Medicine, 28*, 839–855.

Stephens, M. A., Mahon, P. B., McCaul, M. E., & Wand, G. S. (2016). Hypothalamic-pituitary-adrenal axis response to acute psychosocialstress: effects of biological sex and circulating sex hormones. *Psychoneuroendocrinology, 66*, 47–55.

Szyf, M. (2011). DNA methylation, the early-life social environment and behavioral disorders. *Journal of Neurodevelopmental Disorders, 3*, 238–249.

Thorell, L. -H. (2009). Valid electrodermal hyporeactivity for depressive suicidal propensity offers links to cognitive theory. *Acta Psychiatrica Scandinavica, 119*, 338–349.

Tidemalm, D., Runeson, B., Waern, M., Frisell, T., Carlström, E., Lichtenstein, P., & Långström, N. (2011). Familial clustering of suicide risk: a total population study of 11.4 million individuals. *Psychological Medicine, 41*, 2527–2534.

Torgersen, S. (2005). Behavioral genetics of personality. *Current Psychiatry Reports, 7*, 51–56.

Tremblay, R. E. (2010). Developmental origins of disruptive behavior problems: the 'original sin' hypothesis, epigenetics and their consequences for prevention. *Journal of Child Psychology and Psychiatry, 51*, 341–367.

Tsuang, M. T. (1983). Risk of suicide in the relatives of schizophrenics, manics, depressives and controls. *Journal of Clinical Psychiatry, 44*, 396–400.

Turecki, G. (2001). Suicidal behavior: is there a genetic predisposition? *Bipolar Disorders, 3*, 335–349.

Turecki, G. (2014). Epigenetics and suicidal behavior research pathways. *American Journal of Preventive Medicine, 47*, 144–151.

Turecki, G., Ernst, C., Jollant, F., Labonte, B., & Mechawar, N. (2012). The neurobehavioral origins of suicidal behavior. *Trends in Neurosciences, 35*, 14–23.

Van den Bergh, B. R. (2011). Developmental programming of early brain and behavior development and mental health: a conceptual framework. *Developmental Medicine and Child Neurology, 53*(Suppl. 4), 19–23.

van Heeringen, K., Hawton, K., & Williams, J. M. G. (2000). Pathways to suicide: an integrative approach. In K. Hawton, & K. van Heeringen (Eds.), *The international handbook of suicide and attempted suicide* (pp. 224–234). Chichester: John Wiley & Sons.

van Heeringen, K., & Mann, J. J. (2014). The neurobiology of suicide. *The Lancet, 1*, 63–72.

Wasserman, D. (2001). A stress-vulnerability model and the development of the suicidal process. In D. Wasserman (Ed.), *Suicide. An unnecessary death* (pp. 13–27). London: Martin Duniz.

Wasserman, D. (2006). *Depression. The facts*. NY: Oxford University Press.

Wasserman, D. (2016). The suicidal process. In D. Wasserman (Ed.), *Suicide. An unnecessary death* (2nd ed., pp. 27–37). NY: Oxford University Press.

Wasserman, D., & Sokolowski, M. (2016). Stress-vulnerability model of suicidal behaviours. In D. Wasserman (Ed.), *Suicide. An unnecessary death* (2nd ed., pp. 27–37). NY: Oxford University Press.

Wasserman, D., Geijer, T., Rozanov, V., & Wasserman, J. (2005). Suicide attempt and basic mechanisms in neural conduction: relationships to the SCN8A and VAMP4 genes. *American Journal of Medical Genetics Part B: Neuropsychiatric Genetics, 133B*(1), 116–119.

Wasserman, D., Geijer, T., Sokolowski, M., Frisch, A., Michaelowsky, E., Weizman, A., Rozanov, V., & Wasserman, J. (2007a). Association of the serotonin transporter promoter polymorphism with suicide attempters with a high medical damage. *European Neuropsychopharmacology, 17*, 230–233.

Wasserman, D., Geijer, T., Sokolowski, M., Rozanov, V., & Wasserman, J. (2007b). Nature and nurture in suicidal behavior, the role of genetics: some novel findings concerning personality traits and neural conduction. *Physiology and Behavior, 92*, 245–249.

Wasserman, D., Sokolowski, M., Rozanov, V., & Wasserman, J. (2008). The CRHR1 gene: a marker for suicidality in depressed males exposed to low stress. *Genes, Brain and Behavior, 7*, 14–19.

Wasserman, D., Sokolowski, M., Wasserman, J., & Rujescu, D. (2009). Neurobiology and genetics of sucide. In D. Wasserman, & C. Wasserman (Eds.), *Oxford textbook on suicidology and suicide prevention* (pp. 165–182). NY: Oxford University Press.

Wasserman, D., Carli, V., Wasserman, C., Apter, A., Balazs, J., Bobes, J., Bracale, R., Brunner, R., Bursztein-Lipsicas, C., Corcoran, P., Cosman, D., Durkee, T., Feldman, D., Gadoros, J., Guillemin, F., Haring, C., Kahn, J. P., Kaess, M., Keeley, H., Marusic, D., Nemes, B., Postuvan, V., Reiter-Theil, S., Resch, F., Sáiz, P., Sarchiapone, M., Sisask, M., Varnik, A., & Hoven, C. W. (2010). Saving and empowering young lives in Europe (SEYLE): a randomized controlled trial. *BMC Public Health, 10*, 192–207.

WHO (2010). *International statistical classification of diseases and related health problems*, vol. 1–3 (Tenth Revision). Geneva.

WHO (2013). *Suicide data. Geneva.*

WHO (2014a). *Child maltreatment. Fact sheet No. 150.* Geneva.

WHO (2014b). *Preventing suicide: A global imperative.* Executive Summary. Geneva.

Williams, J. M. G., & Pollock, L. R. (2000). The psychology of suicidal behavior. In K. Hawton, & K. van Heeringen (Eds.), *The international handbook of suicide and attempted suicide* (pp. 79–93). Chichester: John Wiley & Sons.

Woodward, J. (1998). *The lone twin: a study in bereavement and loss.* London: Free Association Books.

Zalzman, G., Frisch, A., Apter, A., & Weizman, A. (2002). Genetics of suicidal behavior: candidate association genetic approach. *Israeli Journal of Psychiatry and Related Sciences, 39*, 252–261.

Ideas for Prevention

All studies and models that are trying to explain suicide have an ultimate goal—to suggest approaches, strategies, and methods of prevention. So far as there is a growing understanding of the role of stress and epigenetics in suicide, it would be very relevant to explore how these new insights may help to suggest solutions for the problem of suicides in young people. There is not much hope that the level of psychosocial stress in modern societies will diminish; on the contrary, growing competition and conflicts, political instability in many countries, terrorism, and migration crisis in the world together with mass media coverage of these events increase the pressure and promote more fear and anxiety. It may have an impact on suicides in general and adolescents' suicides in particular. Therefore, discussion and efforts should be focused on the search of ways and approaches of enhancement of internal protective potential of new generations, generally understood as resilience. This may suggest new ideas and approaches for the design of suicide prevention measures.

Suicide prevention in adolescents is a priority in many countries on all continents. Current psychiatric and psychological resources contain a great number of comprehensive original papers and reviews that address this subject. Main directions and approaches to suicide prevention in youth include different types of treatment of depression, psychosocial interventions (variants of cognitive-behavioral therapy [CBT] or other psychotherapy), and pharmacological interventions (Miller, Eckert, & Mazza, 2009; Brent, 2009; Spirito & Esposito-Smythers, 2009). Main arenas for interventions are medical wards, families, schools, communities, and wider societies. School-based suicide prevention initiatives are considered rather promising so far as they can reach targeted contingents and involve them in a preventive activity. Although there are many variants, main approaches are rather limited and comparative effectiveness is rarely evaluated. Recently, a large-scale cluster-randomized controlled SEYLE study enrolling more than 11,000 adolescents from 10 European countries has evaluated the preventive potential of several types of school-based interventions. They included (1) gatekeeper training module targeting teachers and other

Stress and Epigenetics in Suicide. http://dx.doi.org/10.1016/B978-0-12-805199-3.00006-3

school personnel, (2) educational program targeting pupils and aimed to enhance awareness regarding mental health issues, and (3) screening by professionals with referral of at-risk pupils to mental health providers. Educational approach (implemented in a highly interactional manner, with the participation of young trained instructors and utilizing role plays and discussions) proved to be effective both in lowering suicidal ideation and incidence of suicide attempts in 12 months of follow-up (Wasserman et al., 2015). This study is one of the first large-scale studies that compared different strategies for suicide prevention in young people at schools. It is remarkable that intervention aimed at enhancement of pupils' internal potential for overcoming difficulties and avoiding risky behaviors appeared to be most effective. Thus, raising awareness regarding stress, its consequences, providing peer support in stressful and crisis situations, as well as in detection of anxiety and depressive symptoms, and promoting healthy behaviors in young people have a great potential.

Different other innovative approaches based on culture-specific strategies and locally designed interventions, including those using the Internet and mobile phones technologies, involving parents, wide public, communities, voluntary organizations, medical, and other resources are widely represented in different countries (Malone & Yap, 2009). Topics that are discussed and skills trained in youngsters are focused on the understanding of the nature of stress, signs of depression, communication skills, problem solving, bullying and victimization prevention, and healthy life-styles promotion. On the community level, widely recognized topics that are usually addressed are domestic violence, alcohol and substance abuse, unemployment, and poverty (Malone & Yap, 2009). This short overview, absolutely not complete so far as a substantial number of governmental programs and local initiatives have many interesting specific components, which we are unable to present here, gives an impression of vast activity, and many evaluation reports confirm the effectiveness of these measures being rather enthusiastic. Nevertheless, despite many programs and actions the situation still remains serious, and this is acknowledged even in the most economically developed countries where prevention programs are implemented on the regular basis and on the national level. All this leaves an impression of unused resources, unexplored opportunities, and a necessity to suggest new and fresh ideas that may have wider outcomes. Here we will try to present evidence or at least considerations how new fascinating knowledge on the epigenetic programming of health and disease may be helpful in this domain.

CONCEPTUAL FRAMEWORK AND PREVENTION STRATEGIES LIMITATIONS

Many classical suicide prevention strategies are tailored within the narrow domain and are targeted at high-risk groups, which should be first identified. This includes training of medical personnel, school stuff, possible other support providers, or individual treatment of high-risk patients. On the contrary, there is a well-founded and grounded point of view that limitations in existing prevention approaches are based on our perceptions regarding the problem of suicide in young people. Human mind looks for more or less logical explanation of suicide of a very young people to defend one's internal feeling of integrity and psychological stability. As a result, the problem appears in a modified form, especially after scientific analysis because it is very acceptable to attribute it as the result of mental disorder or immaturity of emotional and cognitive processes. On the contrary, the variability of motives, complexity of internal emotions, feelings, and existential problems (actually the level of psycho-emotional or perceived stress) of the young people largely remain unappreciated by adults who are planning suicide prevention. The author of the discussion paper *Youth Suicide as a "Wild" Problem: Implications for Prevention Practice* Jennifer White draws attention to the fact that youth suicide has largely been constructed as a "tame problem," and this, in turn, places certain limits on what might be thought, said, or done in response (White, 2012). Older adults are building their activity on the basis of their perceptions of this problem as tame, that is, logical, stable, and certain. In contrast, youth suicide might be more fruitfully understood as "wild" problem—an unruly problem that is associated with high levels of instability, uncertainty, unpredictability, and complexity (White, 2012). Youth suicide is made a "tame" problem due to psychiatric (and medical in general) thinking, first because in most cases suicide of a young person is represented as a result of mental disorder, though not always diagnosed, and second by "biologification" of suicide and profoundly discussing "suicidal brain," neurotransmitters, genetics, and so on (White, 2012).

Criticism of medical approach is also presented in other publications. For instance, Shahtahmasebi (2008) points that medical model of suicide turns into an attempt to treat depression without addressing suicide. It is not only about the prescription of antidepressant medication to young children, but also about evidence-based methodological resources regarding suicide prevention which often suggest rather an authoritarian approach in interactions with young people to check their psychological well-being. As a result, the whole system of prevention may become part of the problem rather than the solution (Shahtahmasebi, 2008). Treatment of depression often is supposed to be relevant for ethnic minority youth and representatives of autochthonous nations. Of course, high prevalence of mental health disorders, as well as increased substance and alcohol abuse, should be taken into consideration, and elimination of these problems may lower suicides. On the contrary, loss of family links and community support due to accepting global cultural norms and cultural clashes with parents, socioeconomic difficulties, marginalization, racism, loss of religious affiliation, inequalities in education, thwarted hopes, and lack of belonging may be more important factors (McKenzie, Serfaty, & Crawford, 2003). These multiple factors may, of course, interact with medical risk factors like alcohol and drugs abuse or depression, but the variability, unpredictability, uncertainty, and instability that are inherent to all above-mentioned factors, which in this or other way determine, predict or trigger indigenous youth suicide makes the problem really "wild." The same can be said about young people in general so far as cultural instability and unification due to globalization are covering most of countries and continents and different nations appear in a rather similar situation. Thus, problems in young generation are much wider, while mental disorders that may emerge are only one of the consequences of the general development. Of course, treatment of disorders in cases, when they reach clinical level, is necessary, though here also problems are evident. There are many reports that support psychological treatment of depression as a relevant method of suicide prevention, but analysis of many publications shows that it is actually unclear whether psychological treatments are more effective than no treatment since no-treatment control groups are very rare in the studies. There is evidence suggesting that CBT interventions may lower suicidality with moderate effect; however, it is unclear whether these effects are sustained (Devenish, Berk, & Lewis, 2016). Moreover, it is interesting that objective analysis trying to link together suicide rates and mental health system indicators showed that countries with better psychiatric services experience higher suicide rates (Rajkumar, Brinda, Duba, Thangadural, & Jacob, 2013). Thus, preventing suicide in adolescents should take into account wide range of reasons and mechanisms of suicide and should encounter all types of interactions between biological, psychological, and social factors and maybe go further than identification of high-risk group and targeted treatment. We would suggest that wider interventions are needed that will cover all youngsters in such a manner that risks will be lowered for the whole group, inevitably embracing the most vulnerable ones.

If one adopts this point of view, it is easier to understand disappointing results of some studies that are promoting gatekeepers training and knowledge enhancement about suicide in youngsters (Labouliere, Tarquini, Totura, Kutash, & Karver, 2015; Ghoncheh, Gould, Twisk, Kerkhof, & Koot, 2016). What may be the outcome of better knowledge if there is still a gap between those who are bearing the knowledge and young people who should be taken care of, but who are so far from an older generation. No surprise that some authors advocate for more engaging and interactive training to ensure knowledge transfer (Labouliere et al., 2015). It again coincides with the opinion of White (2012) who suggests that it would be more useful to respond to this problem with strategies that emphasize multiplicity, interpretation, dialog, and negotiation instead of experts' opinions and didactics certainty. It is more important to ask right questions that to provide right solutions to the young people when it comes about suicides. We would like to add that it is also very important to use all existing knowledge regarding predisposing factors and mechanisms of vulnerability and resilience. It may be said that suicidology appears at the new phase of "biologification" of suicide (and in this case biologification does not bear negative connotation), when social, psychological, and even cognitive processes appear to be factors that are touching biological mechanisms and when biological mechanisms influence cognition, stress reactivity, and emotions related to it. This knowledge can have a bigger impact if wide public will be enlighted; there is a new agenda for objective and persuasive talk about suicide in different auditoriums, both among parents, teachers, and youngsters.

HOW RESILIENCE IS BUILT? THE EARLIER—THE BETTER

Understanding how social and psychological factors, especially perceived stress, can be embedded in brain structures and how it can influence the risk of mental health problems and suicide really opens new opportunities for discussion regarding prevention. What could be suggested on the theoretical and practical level to prevent suicides in the adolescents and young adults on the basis of this knowledge? If we have an explanation of the interrelation and interaction of wide environmental factors (from in utero environment to social environment in the college or school and further in life) with personality (temperament, important traits, and behaviors), cognition (perceived stress, coping abilities), and with biological mechanisms that conserve environmental influences and tend to promote them in generations, can we suggest new approaches that will eventually overcome shortcomings that exist until now? The first and the most evident idea that comes to mind is associated with mechanisms of epigenetic programming of poor mental health and behaviors, conservation of early trauma, and programming of stress vulnerability. If we know conditions which may lead to vulnerable phenotype, we may concentrate not only on measures aimed to eliminate risks that produce vulnerability, but also on the opposite—promoting conditions that may help to develop resilient phenotype.

Resilience is a wide concept that embraces all possible pathways and mechanisms that are associated with the ability to remain safe and sane in a severe stress or crisis. Another side of resilience is the ability to rehabilitate quickly and without serious consequences for mental health after adversity. Of course, one's life can pass without serious negative life events in childhood, but generally speaking, one cannot avoid some of them across the whole life. Speaking about adolescents, many of them experience difficulties and frustrations in rather young age, including serious losses or break of relations, which may contribute to suicidal thoughts and even suicide attempts. On the contrary, very small portion of young people actually commit suicide. For instance, in our study approximately 40% of adolescents reported death phantasies and feeling that life has no meaning, while more definite thoughts about suicide attempt occurred in 33% and plans about attempt occurred in 17%. The attempt itself occurred in 5.4% of adolescents for the last year (Rozanov, Rakhimkulova, & Ukhanova, 2014). As to suicide, existing data for the same region evaluate rate for adolescents as 3 per 100,000, that is, almost 1,800 times less than attempts (Rozanov, Valiev, Zakharov, Zhuzhulenko, & Kryvda, 2012). Barriers that are blocking more severe suicidal behaviors (sometimes referred as "antisuicidal barriers") are complex and predominantly determined by emotional, cognitive and cultural factors. For instance, suicide may be depreciated as a way to resolve traumatic situation or blocking can occur owing to deep beliefs of the sinfulness of suicidal behavior, which are imprinted by religious prohibitions (though the individual may not be a religious personality). It may also be due to social attitudes based on a punishable nature of suicidal acts or due to aesthetic concepts of "ugliness" of suicide, and so on. Thus, antisuicidal barriers are mostly of cultural origin. It means that the level of perceived stress experienced by a suicidal individual and existing vulnerability is counterbalanced by fear, shame, or feeling of sin that is associated with suicide in the consciousness and mentality of the personality. This produces resilience that is, to our mind, based on adherence to certain cultural norms, which are eroded recently due to globalism. So far as our analysis and modeling foresees two types of vulnerable groups among the young population—those that are vulnerable due to early life adversities and have experienced programming effects of stress (structural model, corresponding mismatch hypothesis) and those who are vulnerable due to accumulation of stress during further maturation (functional model, corresponds cumulative stress hypothesis), we will discuss these types of resilience separately. Resilience is a complex and multilayered phenomenon having several dimensions—biological, personal, and social, often represented as psychosocial (Wu et al., 2013). The essence of this concept is often represented in reasoning that among people who experience life-threatening events (for instance, soldiers at war or victims of the natural catastrophe) majority do not develop PTSD, though a substantial portion does (in case of being at war this portion may reach 20–25%). Resilience sometimes is understood as specific behavioral, emotional, and cognitive pattern that leads to "bending and not breaking" or even "active resistance" to adversity through coping mechanisms (McEwen, Gray, & Nasca, 2015). Brain plasticity, especially in the early life periods, and programming of stress-reactivity systems, mainly HPA, are thought to be main underlying factors of such trait or ability. Although studies of resilience are still at an early stage, recent investigations are trying to identify genetic, epigenetic, developmental, psychological, and neurochemical factors that underlie resilience, especially in development. There may exist predispositions to resilience, though it is largely the result of development and, sometimes, even of training. Several systems of the brain, including noradrenergic, dopaminergic, and serotoninergic, glutamate/GABA system, HPA system, system of neurotrophins (BDNF), and some other mediators of emotions and behavioral responses represented by their specific genetic background factors and supported by neural circuitry are thought to be main biological mechanisms for resilience (Wu et al., 2013). Reversible structural changes or even their absence after stress in such brain regions as hippocampus, amygdala, and prefrontal and orbitofrontal cortex in animal studies are associated with better behavioral, emotional, and cognitive outcomes after severe stress, suggesting that the same structures that are responsible for vulnerability are also involved in resilience (McEwen et al., 2015). Among neurophysiological processes that may be involved in the ability to remain active and maintain resilient character traits and adaptive social responses to stress, besides HPA and SAS, neural circuits for reward and fear are discussed. Well-balanced HPA, efficient control from the side of hippocampus, more precise differentiation between dangerous and not significant threats, quick soothing of noradrenergic structures, endurable reward systems that maintain optimism and hope, reasonable modulation of amygdala activity, and effective control from PFC that helps to suppress fear—all are linked to resilience (Southwick & Charney, 2012; Wu et al., 2013). All these peculiarities may be caused by different factors—genetic predispositions, epigenetic programming, and further in life enhancement of the predisposed and programmed functions or behaviors. In one of the studies this is conceptualized as a three-hit mechanism of resilience (Daskalakis, Bagot, Parker, Vinkers, & de Kloet, 2013), which largely corresponds our and other authors' models, but regarding vulnerability (Chapter 5).

So far as PTSD, as well as depression, anxiety, and suicide, is known to be associated with stress and is dependent on $G \times E \times D$ interaction with active participation of epigenetic mechanisms that are involved in early life programming of vulnerability, it is interesting whether low stress in the childhood (or some specific level of stress) may enhance resilience and diminish the probability of suicide. From the perspectives of DOB-HaD hypothesis, early life environment may have a decisive influence on the further development of behaviors and mental disorders, but the main question is what critical features of the early environment are important for resilience. It is well known that H. Seyle has conceptualized several decades ago the existence of positive (eustress) and negative (distress) types of stress. It is also known that each moderate or minor stressful event, if passed with effective coping, gives a feeling of well-being and promotes mental health and self-esteem and is associated with immune system activation (Sapolsky, 2004). McEwen (2016) suggests classifying stress as "good," "tolerable," and "toxic," and argues that even some adverse outcomes may be beneficial if an individual has positive and resilient biological structures. In general, child during rearing and development needs to be subjected to moderate stressors that would trigger stress response systems and keep them in shape, while severe toxic stress may have a deleterious effect. It is also important to encounter parental support as the factor that counterbalances stress reaction in a child.

The earliest periods of life are known to be crucial for programming of the reactivity and responsivity of HPA axis. As it was already presented in Chapter 2, many rodent studies have led to very impressive discoveries regarding the role of maternal care, negative and positive outcomes of brief or prolonged periods of maternal separation, and natural variation in maternal licking–grooming behavior. On the contrary, overwhelming majority of these studies were aimed to investigate negative outcomes of stress and their mechanisms, while resilient animals were not paid so much attention. This tendency has changed, and now resilient outcomes (inability to achieve anticipated pathological state in cases of stress) are the subject of investigations. Quite recently, Romeo (2015) has overviewed these studies from the point of view of beneficial, in contrast to deleterious, effects of mother–pups interactions. Except well-known experimental results that prove lowered stress response in animals that have received high-quality maternal care or in animals that were separated from dams for short periods of time, he points on sex specificity of these effects. It seems that male offspring benefit more from maternal care with regards to future responsivity to stress (Romeo, 2015). This sex specificity is a new aspect of early life programming that needs further investigation and may lead to better understanding of differences in suicidal behavior between males and females.

Very interesting in depth investigation of resilience is dedicated to individual variability of programming and its genetic background. As it was already mentioned briefly in Chapter 5, the programming role of maternal care and early life stress is modulated by the genetic make-up of the organism. Thus, genes are responsible not only for predisposed traits but also for the ability of the organism to react to environmental stimuli. It complies with the provisions of behavioral genetics, which support that not a trait itself is heritable, but the sensitivity to the environmental stimuli, the norm of reaction. According to differential susceptibility hypothesis (supported by genetic data from the study on adolescents' health determinants), genetic susceptibility to environmental influences, both for positive and negative outcomes, is dependent on the set of "plasticity genes" related to critical neurotransmitter systems and neurodevelopment factors, for instance BDNF (Belsky & Beaver, 2011; Beaver & Belsky, 2012). On the basis of this, a concept of vantage sensitivity was proposed, reflecting variation in response to exclusively positive experiences as a function of individual endogenous characteristics (Pluess & Belsky, 2013).

These considerations and all accumulated data on the epigenetic programming of stress reactivity by maternal behavior attract much more attention to parenting styles. Parenting has always been in focus as a determinant of child development; however its importance grows in view of new data. Many authors draw attention to the positive potential of parenting and the possibility to program life of a child through epigenetic mechanisms (Wu et al., 2013; McEwen et al., 2015; Kanherkar, Bhatia-Dey, & Csoka, 2014; Fine & Sung, 2014). Supportive, attentive, and responsible parenting and especially maternal care at the earliest stages of child development may be direct determinants of the biological-based adaptive potential of the child, which will have an impact on many processes within the whole life-span of descendants (Wu et al., 2013). This very practical and actually very clear from the point of view of common wisdom recommendation, nevertheless, is often not followed by many young parents due to different reasons, but mostly due to their own immaturity. In view of this, another relevant recommendation is proposed—to avoid early entry into adult roles, that is, to prevent early pregnancies and births (Wu et al., 2013). For young parents, concentrating on preventive measures and being aware regarding possible hazards may lead to positive instead of negative programming of stress reactivity in children and, subsequently, resilience, better mental health, and lower suicide risk in big contingents. Given that level of psychosocial stress and environmental hazards are not likely to lower in a modern world, more and more efforts are needed on the population level to achieve these goals. Promoting of a conscious attitude to these facts and tendencies in the young people who prepare to become parents may be a solution. It is a vast arena of action for older parents and grandparents, school teachers, medical doctors, psychologists, councilors, and other professionals.

A recent review of such interventions by Gershon and High (2015) provides several examples of community-based measures for parents of newborn and preschool children aged 3–5 years. They include home visits of specialists and teaching parents' positive parenting practices for newborns and home visits of nurses and other councilors together with regular meetings for older children parents. Evaluation has shown that children reared in families that have received additional care and educational support further have demonstrated better school performance, less violent behavior, higher IQ, and lower prevalence of cardiovascular and metabolic disorders (Gershon & High, 2015). These results are easy to interpret and the effect is very encouraging, though the realization of such programs may depend very much on the urban or rural environment, community structure, and the possibility to have access to families. Another interesting intervention is psychotherapy for mothers with depressive symptoms having young children. It appeared that children's functioning improved significantly over time of almost in parallel with improvement in mothers (Swartz et al., 2016).

Early childhood (3–7 years) is a period when new types of stressors (trauma, fears, novelty, and first peer conflicts, aggressive and sexual impulses) emerge, and it is extremely important how these stressors will be counterbalanced by parental support. From the point of view of positive programming and resilience building, for this period of life the concept of "stress inoculation" may be mostly relevant (Levine, 2000; Romeo, 2015). The central idea of inoculation concept is that moderate or short-time intermittent stress experienced in the early life period may prepare the organism to adapt better for future stress, while too much or too little stress may have a negative impact. Animal studies in this field provide quite conclusive data. For instance, in newborn primates, brief intermittent infant stress and short-time maternal separation lead to diminished HPA activation in adolescence and adulthood (Parker, Buckmaster, Sundlass, Schatzberg, & Lyons, 2006). In rodents, similar HPA programming with resilient emotional and behavioral outcomes may be achieved by exposure of young animals to reoccurring periods of novelty (Tang, Akers, Reeb, Romeo, & McEwen, 2006). In general, animal models suggest that brief intermittent exposure to stress develops programmed arousal regulation and resilience (Lyons, Parker, & Schatzberg, 2010). There are also interesting data showing that moderate stress exposure may lead to enhancement of social bonding. Rats, exposed to moderate stress by immobilization, displayed more positive social behavior, such as resource sharing and reduced aggression. It was linked to increase in the prosocial hormone oxytocin. But if the immobilized rats were exposed to a stronger stressor (fox odor), prosocial behaviors were lost (Muroy, Long, Kaufer, & Kirby, 2016). In one of the studies it was found that mice, which appeared resilient to bullying (i.e., were exposed to social defeat protocol but did not develop anhedonia and other behavioral signs of social defeat), showed a greater number of gene expression changes within the mesolimbic dopamine circuits than traumatized animals (Krishnan et al., 2007). There is also a great potential for the search of epigenetic markers of resilience that may be involved in the prevention of addictions (Cadet, 2016).

Thus, stress inoculation is a quickly developing field of study, which adds objective evidence to well-known general wisdom. Avoiding exposure to uncontrollable stress and trauma and experiences of moderate stress controlled by parents may be an important factor of resilient development of a child. In the early life, such mastery can contribute to stress inoculation with reduced overall reactivity to future stressors and challenges. It is proposed that comprehensive classes in effective parenting based on the bonding and attachment knowledge might help to provide a resilience-promoting child-rearing environment and to reduce transgenerational transmission of stress vulnerability (Southwick & Charney, 2012). In a majority of human families or mother–infant dyads, such ability to balance between freedom and being close to the mother, or between exposure to novelty and support, is achieved in a natural way as it was revealed by J. Bowlby and M. Ainsworth while studying traditional ethnic groups. Development of secure attachment is the function of many variables including child and parent temperament and, in general, is the sign of the potential for positive programming. On the contrary, in modern urban situations with work-related stress and information overload, additional knowledge and training are needed for young parents to summon the mastery of parenting. Positive results from parents training intervention suggest success, though depths of training and knowledge may vary.

The role of moderate stress as a protective factor for future stress reactivity is confirmed in children exposed to normal everyday stress (Tronick, 2006) and in adopted children (Gunnar et al., 2009). These and related investigations of infants confirm the hypothesis that behavioral and physiological resilience develops in part from infants' and young children's experience coping with the normal stress of daily life and social interactions. Moreover, in concordance with the evolutionary-developmental theory study of preschool children showed that exposure to either highly protective or acutely stressful environments results in heightened stress reactivity several years later when children enter the school environment, while moderate stress results in the lowest reactivity levels (Ellis, Essex, & Boyce, 2005).

Therefore, such factors as loving and supportive environment (in a wide sense—family, community, society) and on a personal level (warm, balanced and attentive parenting, good mother care and balanced emotional state of the mother, healthy and joyful mothering) together with positive early life experiences and secure attachment with mother (good and reasonable balance between novelty exposures, independence and support, moderate stress and rewarding activities, understanding of the nature of hyperprotection or neglect and disorganization) constitute positive background that may help to develop resilient and healthy child that will move on to further and much more demanding environments like school, sports team, or peer group. Such basic resilience is very important; however, with approaching adolescence that is associated with fast biological changes and new goals and challenges, more effort is needed to maintain resilient maturation. Furthermore, we will try to discuss possible ideas and practices that may ensure sustaining of positive development during adolescence.

ADOLESCENTS' RESILIENCE—THE POWER OF BODY AND MIND

Adolescents and young adults in the course of their maturation become more dependent on their relations with parents, peers, and mentors, more influenced by emotions, more dependent on self-esteem, physical health, more often experiencing feelings of shame or guilt, and so on. Therefore, the number of stressful influences and varying emotions is growing, which implies the need of developing of traits that may be protective against distress, anxiety, depression and may have potential for attenuating risky and allostatic behaviors, and maladaptive cognition, that is, main predisposing factors of suicide. Analysis made by D. Wasserman in her stress vulnerability and suicidal process model (Fig. 5.2) suggests that main factors that are protective in terms of suicidal activity belong to four main domains: (1) cognitive style and personality, (2) family patterns, (3) cultural and social factors, and (4) environmental factors. Within the first domain such traits as the sense of personal value and confidence in oneself from one side, and seeking help and advice associated with openness to others' opinions and new knowledge from another side, are mentioned. The protective family pattern is represented by general positive, warm, and supportive relations within the family. Cultural and social factors include adherence to specific cultural values and traditions, positive and supportive relationships, social inclusion (groups, friends, sports, church, etc.), and sense of purpose with one's life. Among environmental factors, sleep, diet, light, exercising, and healthy life-styles are mentioned (Wasserman, 2001).

Southwick and Charney (2012) when discussing psychosocial factors of stress resilience and capacity to resist depression include such traits as positive emotions and optimism, having loving caretakers, having role models through life, having a history of mastering challenges, cognitive flexibility and ability to cognitively reframe adversity in a more positive light, and the ability to regulate emotions. They also mention coping self-efficacy, social support, focus on skills development, altruism, commitment to a valued cause or purpose, capacity to extract meaning from adverse situations, support from religion and spirituality, healthy life-styles with a lot of physical activity, and the capacity to rapidly recover from stress (Southwick & Charney, 2012).

It is therefore very probable that some of the mentioned factors may have more profound biological effects including triggering epigenetic mechanisms and programming resilience in the manner similar to programming vulnerability. In one of the last reviews on epigenetics, the authors have outlined the main vector of different environmental influences in shaping health and well-being across the life-span. Such factors as exercise, microbiome, and alternative medicine are known to leave beneficial marks, while toxic chemicals, and drugs of abuse producing harmful marks. Effects of diet, seasonal changes, psychological state, financial status, social interactions, therapeutic drugs, and disease exposure may be both harmful and beneficial depending on the specific nature of the influence (Kanherkar et al., 2014). It is noticeable that many factors that are supposed to be promoting stress resilience and are protective regarding depression and suicide actually overlap. Accumulating data that chemical, psychological, and social factors can trigger epigenetic events in our body and by modulating immune and neuroendocrine system and the brain can actually shape our physical and mental health is both alarming and encouraging. It is therefore extremely important from a theoretical and practical point of view to evaluate the role of biological (and when available—epigenetic) correlates of protective and positive influences that underlie resilience.

Adherence to healthy life-styles, good mental and physical health, and psychological well-being, social integration and inclusion have a tendency to clustering in the same manner as negative tendencies, which also tend to go together. However, among all mentioned positive and protective factors there are few that attract the biggest attention. They may be understood within the "mind–body" paradigm. Since the times of Hippocrates and Galen, body, mind, and soul were perceived as inseparable entities. It was Descartes who introduced a dualistic model, and for centuries materialism and positivistic science have been concentrating on bodily mechanisms, while mind and psyche were excluded from positivistic analysis. However, modern science, by exploring such fields as psychoimmunology, stress, biological effects of psychotherapy, spirituality, and mental practices, is bridging body and mind again (Rao, 2004). It seems that epigenetics may be part of this bridge.

There is a wealth of data that physical exercise has a beneficial effect as a factor of stress resilience, positive mental health, and indirectly may serve as protection factor for suicide (McEwen et al., 2015; Southwick & Charney, 2012; Wasserman, 2001). One of the studies, which compared problem-solving psychotherapy and physical activity promotion in young people, showed that physical activity was more effective in reducing symptoms of depression and anxiety (Parker et al., 2016). In recent reviews on this topic, several identified biological correlates of positive effects of physical activity are presented. Objective studies provide an evidence that regular exercise affects neurobiological factors of resilience by increasing hippocampal volume, which has positive effects on such components of well-being as balanced mood, lower depression, and higher self-esteem. In animal and human studies, running had antidepressant-like effects in behavioral tests and enhanced neurogenesis in brain structures (Southwick & Charney, 2012; McEwen et al., 2015). Recently, Kanherkar et al. (2014) presented a comprehensive review of currently reported epigenetic effects of physical exercise. Among them are beneficial changes in DNA methylation that emerge directly in muscle tissue, which is quite logical, so far as exercise, especially intensive, provides muscles mass growth and physical strength. For instance, acute exercise is associated with DNA hypomethylation of the entire genome in skeletal muscle cells of sedentary individuals, while high-intensity exercise causes the reduction in promoter methylation of certain genes in healthy subjects. Besides changes in muscles, positive changes in DNA methylation after exercise occur within adipose tissue, which is associated with metabolic effects, and within leukocyte cells, which may have a beneficial effect on the immune system. Overall, opinion is expressed that exercise is the way with which an individual can modify one's epigenome to preserve and prolong life (Kanherkar et al., 2014). Here we can see how general wisdom is confirmed by elucidating internal biological mechanisms, which underlie effect known for centuries.

The link to the brain is also revealed—exercise induces a series of beneficial responses in the brain associated with the increase in BDNF. In particular, β-hydroxybutyrate—a metabolite increased after prolonged exercise—actually induces the activities of BDNF promoter (Sleiman et al., 2016). In confirmation, recently several groups of authors are studying mechanisms by which exercise is modulating stress, inflammation, and growth of a child trying to understand mechanisms by which timely administered physical exercise may have long-lasting consequences for health, development, longevity, and immunity. It has been known for centuries, but now it is becoming increasingly clear that epigenetic mechanisms play a critical role of beneficial effects of physical activity in children. In view of the current epidemic of physical inactivity and obesity in children, authors, as a logical practical step, advocate to optimize environments, for instance, schools and playgrounds, and to enlighten child health professionals providing them with more precise and objective knowledge (Cooper, Nemet, & Galassetti, 2004; Radom-Aizik & Cooper, 2016). It is worth paying attention to recommendation regarding knowledge enhancement—though practically all educated people today may know that exercise is good for a child. When parents are often overwhelmed by the idea that their main goal is development of cognitive abilities of their offspring (feeling intuitively that brain and cognition are becoming a limiting factor in modern information society), the focus on the body and physical exercise is really important. This is a very good topic for suicide prevention as well.

There is one more interesting observation, linked to the previous reasoning, that makes the problem of mind and body more acute recently and which turns us back to very early development of a child. Modern children are developing in much-enriched environments. The amount and refinement of audiovisual content, videogames, Internet resources, movies on TV and in the web, and possibilities of modern smartphones, with virtual reality devices approaching—all this creates a very diverse and entertaining informational and emotional context. From the logic of brain development, such enrichment should be beneficial for the child. Very often parents are happy that their children, who still cannot speak, nevertheless are able to operate smartphones or iPads. On the contrary, development of neural circuits must have a certain consistency, which is based on hundreds thousands of years of typical pattern, in which motor functions, emotions, impressions from the external world, and speech are interacting. It is a very context-dependent and flexible process that allows many variations; however, new reality seems to interfere with this process in a very authoritative way. With regard to this, the opinion of the pediatric specialist that external coupling of the brain to TV, DVDs, iPads, and smartphones cannot replace direct multisensory interaction that is provided by parenting and does not seem to have the same advantageous effects for children under 2 years of age that it may have for older children looks very reasonable (Lagercrantz, 2016).

Epigenetic findings in physical exercise fill the old maxima "healthy mind in healthy body" with new highly technological and convincing contents. On the contrary, the causation may be bidirectional. Not only activity in the muscles can lead to epigenetic events in the critical brain systems, but also mental activity, like memorizing emotional regulation rules and mastering psychological self-practices, can change epigenetic profile, for instance, in immune cells. Psychotherapy (cognitive behavioral and dialectical behavioral psychotherapy), mentalization-based therapy, and mindfulness training, as recent studies show, can trigger biological mechanisms. Already in 2000, Gabbard has reviewed animal and human studies on neuroplasticity and collected numerous examples of specific measurable effects of psychotherapy on the brain, which can lead to modification of implicit memory (Gabbard, 2000). More recently, Riess (2011) has pointed that measurable changes in central and peripheral neurophysiology during psychotherapy are determined by the patient–doctor relationship, which may have neurophysiological correlates. Thus, changing implicit memory, training skills how to suppress negative and pervasive thoughts, building capacity to overcome anxiety, and depressive emotions with the use of specific words and empathic attitudes change brain neurophysiology through neuroplasticity mechanisms, which may be associated with epigenetic events. All this is an example of overcoming Cartesian dualism and an opening of new perspectives in mind–body interactivity and interrelatedness (Gabbard, 2000).

Recent studies provide a confirmation. For instance, psychotherapy (4–6 weeks of CBT and DBT) in borderline personality disorder and in panic disorder (both being risk factors for suicidal behavior) produces measurable changes in methylation of BDNF and MAOA in plasma and leucocytes in subjects responsive to therapy (Perroud et al., 2013; Ziegler et al., 2016). In another study, war veterans with diagnosed PTSD were evaluated for methylation pattern of promoter regions of GR and FKB5 genes in blood prior psychotherapy, after 12 weeks of treatment and in 3 months of follow-up. It was shown that while methylation pattern of GR promoter predicted those who would respond to psychotherapy, the methylation status of FKB5 gene was associated with therapy responders' status. The authors conclude that glucocorticoid-related genes are subject to environmental regulation throughout life and that psychotherapy constitutes another contour of "environmental regulation" that may alter epigenetic state of critical genes associated with stress (Yehuda et al., 2013). Although these results are represented by the authors as preliminary, they give a very important example of the involvement of epigenetics in the mechanism of action of the psychotherapeutic intervention that will definitely lead to new research.

These interesting facts and general understanding of links and causal relations between mind, body, thoughts, genetics, epigenetics, and neurobiology of brain give the right to announce a new arena of studies—psychosocial genomics (Rossi, 2002). As the author argues, psychosocial genomics brings together a variety of interdisciplinary fields, including studies of stress, psychosomatics, psychoneuroendocrinology, psychoimmunology, and psychobiological aspects of creativity, optimal performance, dreaming, art, ritual, culture, and spiritual life (Rossi, 2002). If compared with "human social genomics" concept promoted by S. Cole (Chapter 4), it is remarkable how logically prefix "psycho" complements this concept. In concordance with these ideas, McEwen and coauthors are advocating for "top-down" interventions that involve integrated mental activity as a source of compensation of vulnerabilities acquired due to early life adversities. He includes in this category psychotherapy, physical activity, programs that promote social support, social integration, and developing meaning and purpose in life (McEwen et al., 2015). Wu and coauthors, describing psychosocial factors that enhance resilience, suggest optimism, active coping, high cognitive functioning and autonomy, motivation (sense of the mission) and positive risk-taking, cognitive reappraisal and ability to find bright side in complex situations, social support due to networking and communication, religious belief and having meaning of life, having moral compass and life example, prosocial activity and altruism, humor, and positive thinking (Wu et al., 2013; Southwick & Charney, 2012). These characteristics of the personality are based on values, beliefs, and meanings, which is a vast arena for the positive development of the personality. It is, to our mind, a very important subject with regard to evolution of values and meanings in adolescents for the last decades, as mentioned earlier (Chapter 1). Problems in this sphere are numerous, while solutions are not at all easy and are dependent on wider context.

ADOLESCENTS' RESILIENCE—THE POWER OF MEANING

Epidemiological observations testify that even in the most-developed countries among adolescents paradoxically marked increase in poor decision making is noticed, which may result in accidental death, unprotected sex, self-harm, and increasing experimentation with alcohol, sex, and illegal substances (Fine & Sung, 2014). It is often mentioned that adolescents may very well know what is dangerous for their health and what should not be done, but may not perform according to their better judgment, especially when being influenced by peers. This is thought to be due to lowered cognitive control on reward impulses and other peculiarities of emotions and cognition in this period of life due to misbalances in different critical brain structures maturation. In particular, studies have established that cognitive development in this period is still lagging behind, while the emotional sphere is already well developed, which may result in such outcomes as enhanced risky behaviors, emotional dysregulation, and lack of critical thinking (Fine & Sung, 2014). Thus, besides all possible measures that may compensate such factor as childhood adversities and that are targeted mostly on children, who have suffered different forms of abuse, wider solutions are needed to prevent suicides that are stemming from perceived stress and concomitant risky behaviors, mental health problems, and negative perceptions of the future in adolescents.

Many authors recently are discussing such approaches and interventions. For instance, Fine and Sung (2014) in their discussion on adolescents' brain maturation, McEwen et al. (2015) discussing protective measures aimed to attenuate or even reverse neurodevelopment deficits caused by stress, and authors who are speaking more specifically about suicide prevention in adolescents like Wyman (2014) are suggesting development-oriented and universal strategies. Such strategies are focusing on development assets and resources for children at risk instead of addressing risk lowering. Generally speaking, skills building for life in general may be more important than building skills for risk avoidance. Therefore, activities associated with social interactions, prosocial activity, social support, and finding meaning are suggested as wide strategies that may influence big contingents. Such activities at different stages of life are thought to promote neuroplasticity, building a better basis for future maintenance of cognitive potential even for older age (McEwen, 2016). On the contrary, such activities may have even wider outcomes by promoting intrinsic values and giving substance and meaning to life. The problem of the meaning of life, purpose in life, intrinsic values, self-acceptance, and self-development, to our mind, is of special importance, and we would like to pay more attention to it.

For years having meaning has been considered to play one of the most significant roles in the human ability to feel that one's existence is meaningful and needed, and vice versa, having no meaning has been viewed as a solid basis for a grave crisis, both personal and social. Although specific to human's mental abilities, meaning can have lots of interpretations. Modern researchers tend to concentrate on the connection between the meaning and human needs and differentiate between personal needs and personal meaning and societal needs and societal meanings. For instance, recent analysis of psychological publications has identified such sources of meaning in life as having a significant others, having new experiences, and performing spiritual activities; such components of meaning as focusing on self, connecting to others, contributing to others, and having a sense of direction and purpose; and such emotional outcomes of having meaning in life as happiness, satisfaction, and joy (Noviana, Miyazaki, & Ishimaru, 2016).

Whatever the definition would appeal to a prospector, there is a consensus on the protective role having meaning in life has to be a human being as it helps the person to be at peace with oneself, the world, and others (Steger, 2012). When seeing one's life as meaningful, the person is prone to understand oneself and thus develops positive self-identity and self-acceptance. At the same time, having meaning makes the person feel that he/she can understand the world around, which leads to the development of environmental mastery and self-actualization. In such conditions, it is much easier for the person to find his/her fit within the world and build a bunch of positive relationships (Steger, 2012; Brassai, Piko, & Steger, 2011). This interpretation is calling up concepts of positive mental health (Kobau et al., 2011) and salutogenesis with the sense of coherence as a central idea (Antonovsky, 1996).

Numerous studies provide solid proof that being able to find meaning in one's life is strongly associated with a number of positive outcomes, including positive self-identity and higher self-esteem, better health care and higher scores on subjective health and well-being, higher life satisfaction, positive affect, mood, and happiness. For example, Tavernier and Willoughby (2012) found out that the ability to create meaning out of turning points or significant life experiences in adolescents is related to psychological well-being. The results of the longitudinal study showed that adolescents who were to ascribe less sophisticated meaning to their life events tended to develop more diffused self-identity by the age of 23, while those with more sophisticated meaning had an overall higher identity maturity index, as well as higher generativity and optimism (McLean & Pratt, 2006). Similarly, the subjects who developed more positive life story demonstrated better psychological functioning (Banks & Salmon, 2013).

Having meaning in life is associated with more positive psychological background and attitude to personal health. Those who feel they have a purpose in their lives, look after their health better and would rather take preventive measures than wait last minute when it comes down to their health issues (Kim, Strecher, & Ryff, 2014). These results correspond with the findings of another study, which shows how life satisfaction and meaning in life influence well-being in male and particularly in female adolescents (Góngora, 2014). In another research, the presence of meaning in life was found to have a profound effect on illicit drug and sedatives use among males and binge drinking, unsafe sex, lack of exercise, and diet control among females (Brassai et al., 2011).

Having meaning in life has a protective effect on clinical and pre-clinical mental health problems. Thus, depressed individuals compared with controls generated much less specific life goals and gave less specific explanations of why they were not planning on achieving them while being rather specific on the goals they wanted to avoid (Dickson & Moberly, 2013). Mental health patients who showed better results in their recovery processes also appeared to have a better understanding of their life goals and reflected a broader spectrum of life roles (Clarke, Oades, & Crowe, 2012). Purpose in life appeared to be a key factor linked to resilience and recovery from mental disorders in a study of primary care patients with a history of exposure to a range of severe traumatic events (Alim et al., 2008). Finally, empirical studies testify that having meaning and purpose in life is a protective factor against suicidal ideation and behavior, even in cases of severe stress like bullying victimization (Xie, Zou, & Huang, 2012; Henry et al., 2014). Thus, having meaning in life is a strong factor of suicide resiliency (Kleiman & Beaver, 2013).

Meaning in life is associated with such traits and behaviors as spirituality, religiosity, altruism, moral compass, prosocial behavior, and so on. It is rather well established that spirituality and religiosity, in their own turn, have neurobiological correlates. There is an evidence that self-reported spirituality predicts lower suicidal behavior even in the presence of hopelessness and depression in adolescents (Talib & Abdollahi, 2015). In a more objective study, it is reported that in adult offspring of depressed parents with the high familial prevalence of depression, risk was 90% decreased in case religion and spirituality were important for them (Miller et al., 2014). Moreover, real importance of religion and spirituality (but not frequency of church attendance, which is often registered but may, in reality, be a very superficial index) was associated with thicker cortex in different parts of the brain (Miller et al., 2014). In a remarkable study of Raposa, Laws, and Ansell (2015) involving 77 comparatively young adults, ranging from 18 to 44 years old, subjects were asked to report any stressful life events they experienced every day across several domains (e.g., interpersonal, work/education, home, finance, health/accident), and they were also asked to report whether they had been engaged in various helpful behaviors to other people. Helping to others appeared to be an effective strategy for reducing the impact of stress on emotional functioning (Raposa et al., 2015). Similarly, helping others predicted reduced mortality, specifically by buffering the association between stress and mortality (Poulin, Brown, Dillard, & Smith, 2013). A study of school children in Greece showed that higher altruism resulted in lower classroom competitiveness and was associated with higher empathy and resilience. On the basis of this, authors advocate training of altruism in classes (Leontopoulou, 2010).

It is intuitively clear that meaning in life, prosocial activity, altruism, religiosity, and spirituality are linked together and may represent different facets of each other. It also has a direct relation to values and life goals, and in this sense may be interrelated with human psychological well-being, happiness, and life satisfaction. Very interestingly, in a study of a psychological well-being Fredrickson et al. (2013) have revealed the specific association of conserved transcriptional response to adversity (CTRA) with components of well-being. CTRA is associated with uniform leucocytes basal gene expression profiles under chronic psychosocial stress and is seen as one of the reasons of autoimmune disorders and lowered antiviral immunity in modern humanity (see Chapter 4). It was found that while hedonic well-being was associated with upregulation of CTRA, eudemonic well-being was linked to downregulation of this transcriptional response (Fredrickson et al., 2013). It is a remarkable study so far as hedonic well-being represents the orientation of the individual on self-gratification (extrinsic values), while eudemonic well-being is a deeper feeling associated with striving toward meaning, prosocial activity, and generally well-lived life (intrinsic values) (Ryff & Singer, 1996). It is even more impressive so far as general well-being and depressive symptoms in studied contingent were correlated with subcomponents of well-being rather uniformly, while transcriptional profile appeared to be so sensitive to deep psychological motives and their emotional maintenance.

It is not easy to suggest direct measures or strategies that may promote deeper meaning in the feeling of youngsters' life. More or less conscious understanding of it usually comes in older age. Nevertheless, the situation is not hopeless. Moreover, it is adolescence when such meanings can be developed in a very natural way using something like imprinting mechanism. It is intuitively understood by many parents and communities, which promote positive thinking, making records of positive events and altruistic gestures and doings, more contacts with natural environments, physical activity, helping others, taking care about younger and so on. As it is mentioned by Huppert (2004), "the beauty of interventions that nurture positive emotions, attitudes, and behaviors, is that they can potentially benefit the normal majority of people whose lives may not be as happy or fulfilled as they might wish. This contrasts with the usual approach of restricting interventions to the small minority of the population who already have a problem or are at high risk of developing a problem." On the contrary, it may be suggested that most substantial results may be achieved on the basis of vast social and even political changes that will tackle such items as inequalities, injustice, social exclusion, discrimination, and other social problems. Eliminating such problems wherever possible together with the support of positive parenting and nurturing, accustoming to meaningful activities, productive work, and prosocial actions must give better results. This may be the way of overcoming stress and its epigenetic consequences, as well as building resilience, also very probably with the involvement of epigenetics.

References

Alim, T. N., Feder, A., Graves, R. E., Wang, Y., Weaver, J., Westphal, M., Alonso, A., Aigbogun, N. U., Smith, B. W., Doucette, J. T., Mellman, T. A., Lawson, W. B., & Charney, D. S. (2008). Trauma, resilience, and recovery in a high-risk African-American population. *American Journal of Psychiatry, 165*, 1566–1575.

Antonovsky, A. (1996). The salutogenic model as a theory to guide health promotion. *Health Promotion International, 11*, 11–18.

Banks, M. V., & Salmon, K. (2013). Reasoning about the self in positive and negative ways: relationship to psychological functioning in young adulthood. *Memory, 21*, 10–26.

Beaver, K. M., & Belsky, J. (2012). Gene-environment interaction and the intergenerational transmission of parenting: testing the differential-susceptibility hypothesis. *Psychiatric Quarterly, 83*, 29–40.

Belsky, J., & Beaver, K. M. (2011). Cumulative-genetic plasticity, parenting and adolescent self-regulation. *Journal of Child Psychology and Psychiatry, 52*, 619–626.

Brassai, L., Piko, B. F., & Steger, M. F. (2011). Meaning in life: is it a protective factor for adolescents' psychological health? *International Journal of Behavioral Medicine, 18*, 44–51.

Brent, D. (2009). Effective treatment for suicidal youth. Pharmacological and psychosocial approaches. In D. Wasserman, & C. Wasserman (Eds.), *Oxford textbook on suicidology and suicide prevention* (pp. 667–676). New York: Oxford University Press.

Cadet, J. L. (2016). Epigenetics of stress, addiction, and resilience: therapeutic implications. *Molecular Neurobiology, 53*, 545–560.

Clarke, S., Oades, L. G., & Crowe, T. P. (2012). Recovery in mental health: a movement towards well-being and meaning in contrast to an avoidance of symptoms. *Psychiatric Rehabilitation Journal, 35*, 297–304.

Cooper, D. M., Nemet, D., & Galassetti, P. (2004). Exercise, stress, and inflammation in the growing child: from the bench to the playground. *Current Opinion in Pediatrics, 16*, 286–292.

Daskalakis, N. P., Bagot, R. C., Parker, K. J., Vinkers, K. H., & de Kloet, E. R. (2013). The three-hit concept of vulnerability and resilience: towards understanding adaptations to early-life adversity outcomes. *Psychoneuroendocrinology, 38*, 1858–1873.

Devenish, B., Berk, L., & Lewis, A. J. (2016). The treatment of suicidality in adolescents by psychosocial interventions for depression: a systematic literature review. *Australian & New Zealand Journal of Psychiatry, 50*(8), 726–740.

Dickson, J. M., & Moberly, N. J. (2013). Reduced specificity of personal goals and explanations for goal attainment in major depression. *PLoS One, 8*(5), e64512.

Ellis, B. J., Essex, M. J., & Boyce, W. T. (2005). Biological sensitivity to context: II. Empirical explorations of an evolutionary-developmental theory. *Developmental Psychopathology, 17*, 303–328.

Fine, J. G., & Sung, C. (2014). Neuroscience of child and adolescent health development. *Journal of Counseling Psychology, 61*, 521–527.

Fredrickson, B. L., Grewen, K. M., Coffey, K. A., Algoe, S. B., Firestine, A. M., Arevalo, J. M. G., Ma, J., & Cole, S. W. (2013). A functional genomic perspective on human well-being. *Proceedings of the National Academy of Sciences, 110*, 13684–13689.

Gabbard, G. O. (2000). A neurobiologically informed perspective on psychotherapy. *British Journal of Psychiatry, 177*, 117–122.

Gershon, N. B., & High, P. C. (2015). Epigenetics and child abuse: modern-day Darwinism—the miraculous ability of the human genome to adapt, and then adapt again. *American Journal of Medical Genetics Part C (Seminars in Medical Genetics), 169C*, 353–360.

Ghoncheh, R., Gould, M. S., Twisk, J. W., Kerkhof, A. J., & Koot, H. M. (2016). Efficacy of adolescent suicide prevention e-learning modules for gatekeepers: a randomized controlled trial. *JMIR Mental Health, 3*(1), e8.

Góngora, V. C. (2014). Satisfaction with life, well-being, and meaning in life as protective factors of eating disorder symptoms and body dissatisfaction in adolescents. *Eating Disorders, 22,* 435–449.

Gunnar, M. R., Frenn, K., Wewerka, S. S., & Van Ryzin, M. J. (2009). Moderate versus severe early life stress: associations with stress reactivity and regulation in 10–12-year-old children. *Psychoneuroendocrinology, 34,* 62–75.

Henry, K. L., Lovegrove, P. J., Steger, M. F., Chen, P. Y., Cigularov, K. P., & Tomazic, R. G. (2014). The potential role of meaning in life in the relationship between bullying victimization and suicidal ideation. *Journal of Youth and Adolescence, 43,* 221–232.

Huppert, F. A. (2004). Well-being: towards an integration of psychology, neurobiology and social science. *Philosophical Transactions of the Royal Society of London, Series B, 359,* 1447–1451.

Kanherkar, R. R., Bhatia-Dey, N., & Csoka, A. B. (2014). Epigenetics across the life span. *Cell and Developmental Biology, 2,* 49.

Kim, E. S., Strecher, V. J., & Ryff, C. D. (2014). Purpose in life and use of preventive health care services. *Proceedings of National Academy of Sciences of the United States of America, 111,* 16331–16336.

Kleiman, E. M., & Beaver, J. K. (2013). A meaningful life is worth living: meaning in life as a suicide resiliency factor. *Psychiatry Research, 210,* 934–939.

Kobau, R., Seligman, M. E. P., Peterson, C., Diener, E., Zack, M., Chapman, D., & Thompson, W. (2011). Mental health promotion in public health: perspectives and strategies from positive psychology. *American Journal of Public Health, 101*(8), e1–9.

Krishnan, V., Han, M. H., Graham, D. L., Berton, O., Renthal, W., Russo, S. J., Laplant, Q., Graham, A., Lutter, M., Lagace, D. C., Ghose, S., Reister, R., Tannous, P., Green, T. A., Neve, R. L., Chakravarty, S., Kumar, A., Eisch, A. J., Self, D. W., Lee, F. S., Tamminga, C. A., Cooper, D. C., Gershenfeld, H. K., & Nestler, E. J. (2007). Molecular adaptations underlying susceptibility and resistance to social defeat in brain reward regions. *Cell, 131,* 391–404.

Labouliere, C. D., Tarquini, S. J., Totura, C. M., Kutash, K., & Karver, M. S. (2015). Revisiting the concept of knowledge. *Crisis, 36,* 274–280.

Lagercrantz, H. (2016). Connecting the brain of the child from synapses to screen-based activity. *Acta Paediatrica, 105,* 352–357.

Leontopoulou, S. (2010). An exploratory study of altruism in Greek children: relations with empathy, resilience and classroom climate. *Psychology, 1,* 377–385.

Levine, S. (2000). Influence of psychological variables on the activity of the hypothalamic-pituitary-adrenal axis. *European Journal of Pharmacology, 405,* 149–160.

Lyons, D. M., Parker, K. J., & Schatzberg, A. F. (2010). Animal models of early life stress: implications for understanding resilience. *Developmental Psychobiology, 52,* 402–410.

Malone, K., & Yap, S. Y. (2009). Innovative psychosocial rehabilitation of suicidal young people. In D. Wasserman, & C. Wasserman (Eds.), *Oxford textbook on suicidology and suicide prevention* (pp. 685–690). New York: Oxford University Press.

McEwen, B. (2016). In pursuit of resilience: stress, epigenetics, and brain plasticity. *Annals of the New York Academy of Sciences, 1373,* 56–64.

McEwen, B., Gray, J. D., & Nasca, C. (2015). Recognizing resilience: learning from the effects of stress on the brain. *Neurobiology of Stress, 1,* 1–11.

McKenzie, K., Serfaty, M., & Crawford, M. (2003). Suicide in ethnic minority groups. *British Journal of Psychiatry, 183,* 100–101.

McLean, K. C., & Pratt, M. W. (2006). Life's little (and big) lessons: identity statuses and meaning-making in the turning point narratives of emerging adults. *Developmental Psychology, 42,* 714–722.

Miller, D., Eckert, T., & Mazza, J. (2009). Suicide prevention programs in the schools: a review and public health perspective. *School Psychology Review, 38,* 168–188.

Miller, L., Bansal, R., Wickramaratne, P., Hao, X., Tenke, C. E., Weissman, M. M., & Peterson, B. (2014). Neuroanatomical correlates of religiosity and spirituality. *JAMA Psychiatry, 71,* 128–135.

Muroy, S. E., Long, K. L., Kaufer, D., & Kirby, E. D. (2016). Moderate stress-induced social bonding and oxytocin signaling are disrupted by predator odor in male rats. *Neuropsychopharmacology, 41,* 2160–2170.

Noviana, U., Miyazaki, M., & Ishimaru, M. (2016). Meaning in life: a conceptual model for disaster nursing practice. *International Journal of Nursing Practice, 22*(Suppl. 1), 65–75.

Parker, K. J., Buckmaster, C. L., Sundlass, K., Schatzberg, A. F., & Lyons, D. M. (2006). Maternal mediation, stress inoculation and the development of neuroendocrine stress resistance in primates. *Proceeding of the National Academy of Sciences of the United States of America, 103,* 3000–3005.

Parker, A. G., Hetrick, S. E., Jorm, A. F., Mackinnon, A. J., McGorry, P. D., Yung, A. R., Scanlan, F., Stephens, J., Baird, S., Moller, B., & Purcell, R. (2016). The effectiveness of simple psychological and physical activity interventions for high prevalence mental health problems in young people: a factorial randomized controlled trial. *Journal of Affective Disorders, 196,* 200–209.

Perroud, N., Salzmann, A., Prada, P., Nicastro, R., Hoeppli, M. E., Furrer, S., Ardu, S., Krejci, I., Karege, F., & Malafosse, A. (2013). Responce to psychotherapy in borderline personality disorder and methylation stauts of the BDNF gene. *Translational Psychiatry, 3,* e207.

Pluess, M., & Belsky, J. (2013). Vantage sensitivity: individual differences in response to positive experiences. *Psychological Bulletin, 139,* 901–916.

Poulin, M. J., Brown, S. L., Dillard, A. J., & Smith, D. M. (2013). Giving to others and the association between stress and mortality. *American Journal of Public Health, 103,* 1649–1655.

Radom-Aizik, S., & Cooper, D. M. (2016). Bridging the gaps: the promise of omics studies in pediatric exercise research. *Pediatric Exercise Science, 28,* 194–201.

Rajkumar, A. P., Brinda, E. M., Duba, A. S., Thangadurai, P., & Jacob, K. S. (2013). National suicide rates and mental health system indicators: an ecological study of 191 countries. *International Journal of Law Psychiatry, 36,* 339–342.

Rao, O. (2004). How the mind hurts and heals the body. *American Psychologist, 59,* 29–40.

Raposa, E. B., Laws, H. B., & Ansell, E. B. (2016). Prosocial behavior mitigates the negative effects of stress in everyday life. *Clinical Psychological Science, 4*(4), 691–698.

Riess, H. (2011). Biomarkers in the psychotherapeutic relationship: the role of physiology, neurobiology, and biological correlates of E.M.P.A.T.H.Y. *Harvard Reviews of Psychiatry, 19,* 162–174.

Romeo, R. D. (2015). Perspectives of stress resilience and adolescent neurobehavioral function. *Neurobiology of Stress, 1,* 128–133.

Rossi, E. L. (2002). Psychosocial genomics: gene expression, neuroegenesis, and human experiences in mind-body medicine. *Advances in Mind-Body Medicine, 18,* 22–30.

Rozanov, V. A., Valiev, V. V., Zakharov, S. E., Zhuzhulenko, P. N., & Kryvda, G. F. (2012). Suitsidy I suitsidalnie popytki sredi detey i podrostkov v Odesse v 2002–2010 gg (Suicides and suicide attempts among children and adolescents in Odessa in 2002–2010). *Zhurnal psychiatrii i meditisinskoy psychologii (Journal of Psychiatry and Medical Psychology), 1,* 53–61.

Rozanov, V. A., Rakhimkulova, A. V., & Ukhanova, A. I. (2014). Oschuschenie bessmyslennosti suschestvovaniya u podrostkov—sv'yaz s suitsidalnimi proyavleniayami i psihicheskim zdorov'yem ("Life has no meaning" feelings in adolescents—relation to suicidal ideation and attempts and mental health). *Suitsidologia (Suicidology), 5,* 33–40.

Ryff, C. D., & Singer, B. (1996). Psychological well-being: meaning, measurement and implications for psychotherapy research. *Psychotherapy and Psychosomatics, 65,* 14–23.

Sapolsky, R. M. (2004). *Why zebras don't get ulcer. The acclaimed guide to stress, stress-related disease and coping* (3rd ed.). Henry Holt and Co.: NY.

Shahtahmasebi, S. (2008). Suicide research and adolescent suicide trends in New Zealand. *The Scientific World Journal, 8*, 287–302.

Sleiman, S. F., Henry, J., Al-Haddad, R., El Hayek, L., Abou Haidar, E., Stringer, T., Ulja, D., Karuppagounder, S. S., Holson, E. B., Ratan, R. R., Ninan, I., & Chao, M. V. (2016). Exercise promotes the expression of brain derived neurotrophic factor (BDNF) through the action of the ketone body β-hydroxybutyrate. *Elife*, Jun 2; 5. pii: e15092.

Southwick, S. M., & Charney, D. S. (2012). The science of resilience: implications for the prevention and treatment of depression. *Science, 338*, 79–82.

Spirito, A., & Esposito-Smythers, C. (2009). Individual therapy techniques with suicidal adolescents. In D. Wasserman, & C. Wasserman (Eds.), *Oxford textbook on suicidology and suicide prevention* (pp. 677–683). New York: Oxford University Press.

Steger, M. F. (2012). Experiencing meaning in life: optimal functioning at the nexus of spirituality, psychopathology, and well-being. In P. T. P. Wong (Ed.), *The human quest for meaning* (pp. 165–184). New York: Routledge.

Swartz, H. A., Cyranowski, J. M., Cheng, Y., Zuckoff, A., Brent, D. A., Markowitz, J. C., Martin, S., Amole, M. C., Ritchey, F., & Frank, E. (2016). Brief psychotherapy for maternal depression: impact on mothers and children. *Journal of the American Academy of Child & Adolescent Psychiatry, 55*, 495–503.

Talib, M. A., & Abdollahi, A. (2015). Spirituality moderates hopelessness, depression, and suicidal behavior among Malaysian adolescents. *Journal of Religion and Health* doi: 10.1007/s10943-015-0133-3.

Tang, A. C., Akers, K. G., Reeb, B. C., Romeo, R. D., & McEwen, B. S. (2006). Programming social, cognitive and neuroendocrine development by early exposure to novelty. *Proceeding of the National Academy of Sciences of the United States of America, 103*, 15716–15721.

Tavernier, R., & Willoughby, T. (2012). Adolescent turning points: the association between meaning-making and psychological well-being. *Developmental Psychology, 48*, 1058–1068.

Tronick, E. (2006). The inherent stress of normal daily life and social interaction leads to the development of coping and resilience, and variation in resilience in infants and young children: comments on the papers of Suomi and Klebanov & Brooks-Gunn. *Annals of the New York Academy of Sciences, 1094*, 83–104.

Wasserman, D. (2001). A stress-vulnerability model and the development of the suicidal process. In D. Wasserman (Ed.), *Suicide: An unnecessary death* (pp. 13–27). London: Martin Duniz.

Wasserman, D., Hoven, C. W., Wasserman, C., Wall, M., Eisenberg, R., Hadlaczky, G., Kelleher, I., Sarchiapone, M., Apter, A., Balazs, J., Bobes, J., Brunner, R., Corcoran, P., Cosman, D., Guillemin, F., Haring, C., Iosue, M., Kaess, M., Kahn, J. P., Keeley, H., Musa, G. J., Nemes, B., Postuvan, V., Saiz, P., Reiter-Theil, S., Varnik, A., Varnik, P., & Carli, V. (2015). School-based suicide prevention programmes: the SEYLE cluster-randomised, controlled trial. *Lancet, 385*, 1536–1544.

Wyman, P. A. (2014). Developmental approach to prevent adolescent suicides: research pathways to effective upstream preventive interventions. *American Journal of Preventive Medicine, 47*(Suppl 2), S251–256.

White, J. (2012). Youth suicide as a "wild" problem: implications for prevention practice. *Suicidology Online, 3*, 42–50.

Wu, G., Feder, A., Cohen, H., Kim, J. J., Calderon, S., Charney, D. S., & Mathé, A. A. (2013). Understanding resilience. *Frontiers in Behavioral Neuroscience, 7*, 10.

Xie, X., Zou, B., & Huang, Z. (2012). Relationships between suicide attitudes and perception of life purpose and of life in college students. *Nan Fang Yi Ke Da Xue Xue Bao (Journal of Southern Medical University), 32*, 1482–1485 (Article in Chinese).

Yehuda, R., Daskalakis, N. P., Desarnaud, F., Makotkine, I., Lehrner, A. L., Koch, E., Flory, J. D., Buxbaum, J. D., Meaney, M. J., & Bierer, L. M. (2013). Epigenetic Biomarkers as Predictors and Correlates of Symptom Improvement Following Psychotherapy in Combat Veterans with PTSD. *Frontiers in Psychiatry, 4,* 118.

Ziegler, C., Ritcher, J., Mahr, M., Gajewska, A., Schiele, M. A., Gehrmann, A., Schmidt, K., Lecsch, P., Lang, T., Helbig-Lang, S., Paili, P., Kircher, T., Reif, A., Vossbeck-Elsebusch, A. N., Aroit, V., Wittchen, H. -U., Hamm, A. O., Deckert, J., & Domschke, K. (2016). MAOA gene hypomethylation in panic disorder—reversibility of an epigenetic risk pattern by psychotherapy. *Translational Psychiatry, 6,* e773.

Conclusions

Suicide is a phenomenon that in a wider context may be described as self-destructiveness or autoaggression. There is a discussion whether self-destructiveness is a purely human inclination. It is possible to find some equivalents of stress-induced self-harm in animals, but suicide is a purely human phenomenon possibly due to the specific human feeling of guilt. Why these dark sides of the subconscious or unconscious are enhanced by stress? If we concentrate only on neurobiology, we will not find an answer. On the contrary, there are some signs that thoughts, meanings, and perceptions may be traced on the epigenetic level. Self-destruction is a process, which implies emotions, cognitions, and meanings. Meanings may be the first step, which builds the frame (direction) of thinking. If a young individual is fascinated by the motto "Live Fast, Die Young," the internal meaning of a shortened life resource is activated. After that, all stressful events that may arise in the course of life will only add to existing attitude and will support internal certainty that life is not worth living any more. If so, there is not much chance for reframing negative events and perception of them in a positive light or for extracting strength from adversity. On the contrary, romanticizing of suicide may appear and attitudes toward death will receive support. This may lead to a behavioral recursion-young individual will find a group with the same ideas on the web, for this there is no shortage; on the contrary, such groups are rather numerous.

Another strong subconscious feeling that may promote suicide is feeling of injustice. Injustice is a deep and strong feeling, possibly also inherent only to humans that may either destroy a personality or push to a social explosion. Feelings of a traumatized child are also about injustice; it is not abuse itself, but the injustice of the situation is the main toxic factor. Of course, injustice, inequalities, and all that is associated with these feelings always existed in the world, but it has never been so evident and young people have never been so strongly exposed to it. Mass media and ubiquity of Internet content are to blame. Ways and standards of beautiful life, entertainment, and hedonistic rewarding behavior are becoming familiar and luring to people globally inviting them to live the same lives. But it is impossible and many young people vaguely understand it. The result will be frustrations, feeling of inferiority, feeling of being a loser, inevitable depressive thoughts, anxiety, and quite possibly addictions, risky behavior, and so on, all signs of perceived stress. Epigenetic events, if they take place, may be the factor of turning these feelings into a chronic state as well as adding more symptoms. Thus, epigenetics seems to be involved in all environmental effects. Suicide as a multifactorial phenomenon involving existential, psychological, behavioral, cognitive, social, and biological reasons perhaps best reflects the role of interactions and interrelations of these factors. We have tried to develop an integrative model that united these factors on the basis of the central role of stress and bridging role of epigenetic events.

The bright side of epigenetics is that epigenetically driven phenotypes are reversible. Although there are still few objective studies, the potential for evaluation of positive thinking, prosocial activity, and eudemonic personality orientation is great. There are also opinions that there is a potential of reversing epigenetic marks with diet supplements, pharmacological agents, and specially designed epigenetic drugs. Well, why not? It is known that healthy diet is a protective factor, together with light, minimum ecological hazards, and other positive environmental qualities. There is also a great potential in evaluating epigenetic effects of yoga, Tay-Chi, meditation, and other practices that are inherent to Oriental culture which seems to keep more strictly to traditions that existed for centuries. Epigenetics establishes a scientific basis for how external factors and the environment can shape an individual both physically and mentally. Therefore, it would be really fascinating to prove that "we are what we think," not only "what we eat" (this already can be considered proven, as modern nutritional epigenetics says). It would be, in view of new thinking, very interesting to look once more at the societies, communities, ethnics, and nations that historically have low suicide rate and try to understand what may the reason having in mind epigenetic effects.

It may be said that from the practical point of view epigenetics already complemented existing wisdom how to nurture children and how to achieve healthy and peaceful life. On the contrary, epigenetics have done even more-it explained why this wisdom is actually right, and what is really right, gave a scientific explanation. This is another positive outcome of epigenetics. In a modern information society, evidence, knowledge, and support from the science may strengthen positive views and help to promote positive practices in wider circles. It means that educational strategies have a good perspective and that psychologists, coaches, and life-advisors may be inspired.

Stress and Epigenetics in Suicide is an open discussion book; its main message is that though we are so technological today, new horizons opened by epigenetics cannot ultimately lead us to final solution of the problem of youth suicide. Suicide is a too human phenomenon, still unpredictable, and will remain so. Our aim is to draw attention to new conditions in which modern youth is developing, which may have consequences associated with epigenetic transformations and to look for the ways how to improve the situation. We hope at least this aim is achieved.

Index

Printed in the United States
By Bookmasters